An Introduction to Design of Experiments

Also available from ASQ Quality Press

Concepts for R&R Studies
Barrentine, Larry B.

Design & Analysis of Experiments, Fourth Edition
Montgomery, Douglas C.

World Class Quality: Using Design of Experiments to Make It Happen
Bhote, Keki R.

Basic References in Quality Control: Statistical Techniques:
(Sixteen different booklets are available—selected volumes listed below)

Volume 5: How to Run Mixture Experiments for Product Quality, Revised Edition
Cornell, John A.

Volume 8: How to Apply Response Surface Methodology, Revised Edition
Cornell, John A.

Volume 14: How to Construct Fractional Factorial Experiments
Gunst, Richard F. and Mason, Robert L.

Statistical Quality Control Using Excel (with software)
Zimmerman, Steven M. and Icenogle, Marjorie L.

Statistical Procedures for Machine and Process Qualification
Dietrich, Edgar and Schulze, Alfred

Business Process Improvement Toolbox
Andersen, Bjørn

To request a complimentary catalog of publications, call 800–248–1946.

An Introduction to Design of Experiments

A Simplified Approach

Larry B. Barrentine

ASQ Quality Press
Milwaukee, Wisconsin

An Introduction to Design of Experiments: A Simplified Approach
Larry B. Barrentine
8724 Warm Springs Way
Knoxville, TN 37923
(423) 692–9950

Library of Congress Cataloging-in-Publication Data

Barrentine, Larry B., 1938-
    An introduction to design of experiments : a simplified approach /
Larry B. Barrentine.
        p.   cm.
        Includes index.
        ISBN 0-87389-444-8 (alk. paper)
        1. Experimental design.   I. Title.
QA279.B37 1999
001.4'34--dc21   98-42759
                                                                                  CIP

© 1999 by ASQ

All rights reserved. No part of this book may be reproduced in any form or by any means, electronic, mechanical, photocopying, recording, or otherwise, without the prior written permission of the publisher.

10 9 8 7 6 5 4 3 2

ISBN 0-87389-444-8

Acquisitions Editor: Ken Zielske
Project Editor: Annemieke Koudstaal
Production Coordinator: Shawn Dohogne

ASQ Mission: The American Society for Quality advances individual and organizational performance excellence worldwide by providing opportunities for learning, quality improvement, and knowledge exchange.

Attention: Bookstores, Wholesalers, Schools and Corporations:
ASQ Quality Press books, videotapes, audiotapes, and software are available at quantity discounts with bulk purchases for business, educational, or instructional use. For information, please contact ASQ Quality Press at 800-248-1946, or write to ASQ Quality Press, P.O. Box 3005, Milwaukee, WI 53201-3005.

To place orders or to request a free copy of the ASQ Quality Press Publications Catalog, including ASQ membership information, call 800-248-1946. Visit our web site at http://www.asq.org.

Printed in the United States of America

∞ Printed on acid-free paper

Quality Press
611 East Wisconsin Avenue
Milwaukee, Wisconsin 53202
Call toll free 800-248-1946
http://www.asq.org
http://standardsgroup.asq.org

Contents

Preface vii

CHAPTER 1 Introduction 1

CHAPTER 2 Experiments with Two Factors 5
    Example 1: Bond Strength 5
    The Eight Steps for Analysis of Effects 7
    The Analytical Procedure 8
        1. Calculate Effects 8
        2. Make a Pareto Chart of Effects 11
        3. Calculate the Standard Deviation of the Experiment, S_e 11
        4. Calculate the Standard Deviation of the Effects, S_{eff} 13
        5. Determine the t-Statistic 13
        6. Calculate the Decision Limits and Determine the Effects 13
        7. Graph Significant Effects 14
        8. Model the Significant Effects 14
    Review of the Experimental Procedure 16
    Example 2: Water Absorption in Paper Stock 17
        1. Calculate Effects 17
        2. Make a Pareto Chart of Effects 19
        3. Calculate the Standard Deviation of the Experiment, S_e 20
        4. Calculate the Standard Deviation of the Effects, S_{eff} 20
        5. Determine the t-Statistic 20
        6. Calculate the Decision Limits and Determine the Effects 20
        7. Graph Significant Effects 20
        8. Model the Significant Effects 21
    A Different Analytical Technique: The Spreadsheet Approach 21
    Exercise 1: Coal-Processing Yield 23
    Nonlinear Models 23
    Exercise 2: Corrosion Study 25

CHAPTER 3 Experiments with Three Factors: 2^3 27
    Example 3: Chemical-Processing Yield 27
    Variation Analysis 31
    Exercise 3: Ink Transfer 33
    Analysis with Unreplicated Experiments 34

CHAPTER 4 Screening Designs 37

Example 4: An Eight-Run Plackett-Burman Design with Seven Factors 40
Reflection 46
Other Analytical Considerations 49
Exercise 4: Nail Pull Study 50
Twelve-Run Plackett-Burman 51
Example 5: Moldability Analysis 51
Exercise 5: The Lawn Fanatics 55
Even Larger Designs 57
Other Types of Designs 57

CHAPTER 5 Problems, Questions, and Review 59

Problems and Questions 59
Review of the Basics in Managing a DOE 61
What Inhibits Application of DOE? 62
Software 62
Appendix 63
Conclusion 63
References 63

Appendix 65

Table A.1: Table of t-Values for $\alpha = .05$ (95% confidence) 66
Table A.2: F-Table for $\alpha = .10$ 67
Exercise 1: Coal-Processing Yield 68
Exercise 2: Corrosion Study 72
Exercise 3: Ink Transfer 76
Exercise 4: Nail Pull Study 81
Exercise 5: The Lawn Fanatics 85
Tables A.11 and A.12: Analysis Tables for 2^3 Factorial 90–91
Table A.13: Design Matrix for $\frac{1}{2}$ Fractional Factorial for Four Factors 92
Table A.14: Analysis Table for $\frac{1}{2}$ Fractional Factorial with Four Factors 93
Table A.15: Design Matrix for $\frac{1}{4}$ Fractional Factorial for Five Factors 94
Table A.16: Analysis Table for $\frac{1}{4}$ Fractional Factorial with Five Factors 95
Table A.17: Design Table for $\frac{1}{2}$ Fractional Factorial with Five Factors 96
Table A.18: Analysis Table for $\frac{1}{2}$ Fractional Factorial with Five Factors 97
Table A.19: Confounding Pattern for Eight-Run Plackett-Burman Design 98
Table A.20: Confounding Pattern for 16-Run Plackett-Burman Design 98
Table A.21: Analysis Table for Eight-Run Plackett-Burman Design 99
Table A.22: Analysis Table for Reflection of Eight-Run Plackett-Burman Design 100
Table A.23: Analysis Table for Eight-Run Plackett-Burman Design with Reflection 101
Tables A.24 and A.25: Analysis Tables for 12-Run Plackett-Burman Design 102–103
Table A.26: Analysis Table for 12-Run Plackett-Burman Design with Reflection 104
Table A.27: Analysis Table for 16-Run Plackett-Burman Design 105
Table A.28: Analysis Table for 16-Run Plackett-Burman Design with Reflection 106
Table A.29: Design Matrix for 20-Run Plackett-Burman 107

Glossary 109

Preface

This book is intended for people who have either been intimidated in their attempts to learn about Design of Experiments (DOE) or who have not appreciated the potential of that family of tools in their process improvement efforts. This is an introduction to the basics, not a complete coverage of this fascinating field. If more people become familiar with and begin applying the basics, they should be easily encouraged to go on to more advanced materials in DOE. Once anyone has had a success with DOE, he or she rarely needs prodding to continue to learn more about this powerful tool. Every effort has been made to simplify the approach and minimize the complexity of the material. Inevitably, there are some points that might have been covered with more statistical depth. It is the author's strong recommendation that anyone who completes this book immediately continue on to a more in-depth reference (some excellent ones are identified in the conclusion of this book), both to learn about areas that are not covered here and to broaden the reader's depth of knowledge of DOE in general.

I am greatly indebted to Woody Greene, Tom Waters, and Mark Black for their assistance and suggestions. I am also deeply appreciative of the support by Bob Doyle and his staff at Simplex Products. Without their support, this book would not have been possible.

Chapter One

Introduction

The purpose of this introduction to the Design of Experiments (DOE) is to showcase the power and utility of this statistical tool while teaching the audience how to plan and analyze an experiment. It is also an attempt to dispel the conception that DOE is reserved only for those with advanced mathematics training. It will be demonstrated that DOE is primarily a logic tool that can be easily grasped and applied, requiring only basic math skills. While software would make the calculations more painless and provide greater versatility, it is necessary to understand what the software is doing. To this end, software is not used with this text, but calculators are used instead to insure that the basics are learned. At the conclusion, software applications will be obvious and some options among available packages will be described. This is by no means a complete treatment of the broad field of DOE. The intent is to introduce the basics, persuade the reader of the power of this tool, and then recommend resources for further study. The material covered will still be sufficient to support a high proportion of the experiments one may wish to perform.

The prerequisites of this book are familiarity with the concepts of process stability, basic statistical process control (SPC), and measurement analysis. As in any process improvement activity, it is necessary to recognize that a process is made up of input variables, process variables, and output measures (see Figure 1.1). The intent is always to improve the output measure, which is labeled as the *response*. There is no direct control on the response variable; in the classical cause-and-effect approach, it is the effect.

The *causes* are what dictate the response. To control the response, one must control the causes, which may be input variables and/or process variables involving the five elements shown in Figure 1.1. (These variables or causes will later be referred to as *factors*.)

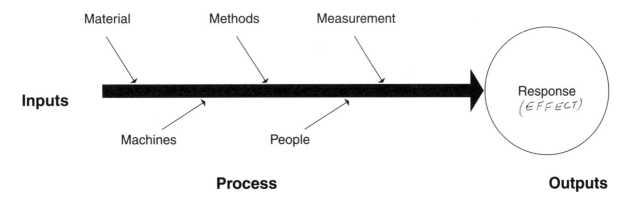

FIGURE 1.1. Cause-and-effect relationship.

For example, there is no control setting in a sales process that allows one to set a sales level. To control sales, one must address those variables that *cause* sales to change, e.g., promotional literature, call frequency, pricing policies, credit policies, personal sales techniques, etc. A process may be very simple, or it may be a complex group of processes.

In concert with this cause-and-effect or systems approach to the process, the concepts of process variation must be understood. Every response demonstrates variation. This variation results from (a) variation in the known input or process variables, (b) variation in the unknown process variables, and/or (c) variation in the measurement of the response variable. The combination of these sources results in the variation of that response. This variation is categorized by the classic SPC tools into two categories: (a) *special-cause variation*—unusual responses compared to previous history; and (b) *inherent variation*—variation that has been demonstrated as typical of that process.

A side note is needed here on terminology. Inherent or typical variation has a variety of labels that are often used interchangeably. In control charting, it is referred to as *common-cause* variation. In control systems, it is called *process noise*. In DOE, it is called *experimental error* or *random variation*. To minimize confusion, it will be referred to in this text as either inherent variation or experimental error.

Control charts are used to identify special-cause variation and, hopefully, to identify the process variables or causes that led to such unusual responses. The presence of special causes within an experiment will create problems in reaching accurate conclusions. For this reason, DOE is more easily performed after the process has been stabilized using SPC tools. The presence of inherent variation also makes it difficult to draw conclusions. (In fact, that is one of the definitions of statistics: decision making in the presence of uncertainty or inherent variation.) If a process variable causes changes in the response that exceed the inherent variation, we state that the change is *significant*.

Inherent variation can also be analyzed to determine if the process will consistently meet a specification. The calculation of process capability is a comparison of the spread of the process with the specifications, resulting in test statistics such as C_p and C_{pk}. Figure 1.2 illustrates the comparison of a process with its upper and lower specification limits.

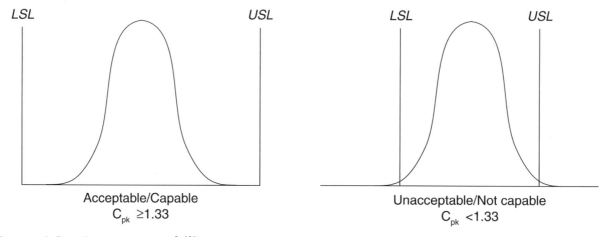

FIGURE 1.2. Process capability.

Design of Experiments is the simultaneous study of several process variables. By combining several variables in one study instead of creating a separate study for each, the amount of testing required will be drastically reduced and greater process understanding will result. This is in direct contrast to the typical *one-factor-at-a-time* approach or *OFAT*, which limits the understanding and wastes data. Additionally, OFAT studies can not be assured of detecting the unique effects of combinations of factors (a condition later to be defined as an *interaction*).

Design of Experiments includes the entire scope of experimentation, including defining the output measure(s) that one desires to improve, the candidate process variables that one will change, procedures for experimentation, actual performance of the experiment, and analysis and interpretation of the results. The objectives of the experimenter in a DOE are

1. to learn how to change a process average in the desired direction
2. to learn how to reduce process variation
3. to learn how to make a process *robust* (i.e., make the response insensitive to uncontrollable changes in the process variables)
4. to learn which variables are important to control and which are not

For ease of instruction, small experiments are presented first, followed by large experiments. In the real world, one would prefer to start with large experiments and progress to smaller ones in order to identify variables that affect the response variables. Terms and definitions are covered as they arise. The terminology used in DOE is often different from the equivalent terms in SPC and is presented to assure that other references are more readable. The initial example is used to define most of the unique terminology and much of the analytical technique.

Chapter Two

Experiments with Two Factors

It is beneficial at this point to review the logic steps to be followed in planning and implementing a DOE, even if some of the terminology has not been defined (this is repeated after the first exercise):

1. Define the process to be studied.
2. Determine the response(s).
3. Determine the measurement precision and accuracy.
4. Generate candidate factors.
5. Determine the levels for the selected factors.
6. Select the experimental design.
7. Have a plan to control extraneous variables.
8. Perform the experiment according to the design.
9. Analyze and draw conclusions.
10. Verify and document the new process.
11. Propose the next study.

An experiment with two factors is presented to start the process of defining terms and procedures. The techniques developed in this small experiment can then be used in experiments with many more factors.

Example 1: Bond Strength

A manufacturer of laminated papers takes two rolls of kraft paper and extrudes a layer of polymer in between, simultaneously pressing the three components into a *sandwich*. The customer has complained about the lack of adherence of the polymer to the two paper layers. Therefore, the objective of the study is to maximize the bond strength. Note that the objective must be defined up front. The manufacturer has developed a test for this bonding using a Bond Meter. (Don't look for this in your equipment catalog!) This is the key output measure; in DOE, it is called the *response variable*. The team working on improving the bond has decided that the best candidate *factors* for controlling bond strength are polymer temperature and paper source. *Factors* are process variables that can be controlled at will during the experiment. Think of them as *knobs* that you can turn as you wish. (Such factors are also referred to as *independent variables* in some references.) These factors were selected based on the review of data from the process and on the brainstorming by the team. After much discussion, the *levels*—the different options or settings for each factor in the experiment—were set as shown below.

Factor	Low Level	High Level
A. Polymer Temperature, °F	580	600
B. Paper Source	Vendor Y	Vendor X

Note that factor A is a *quantitative* factor since its levels can be set along a relatively continuous measurement scale. Factor B is *qualitative* since its levels are discrete; i.e., there are a finite number of levels available. Where possible, quantitative factors are preferred since that permits the experimenter to follow such a factor to an optimum condition with respect to the response.

Next, the *experimental design* must be defined. The experimental design is the definition of the collection of trials to be run in the experiment. In this example, the design is all possible combinations of the chosen factor levels, called a *full-factorial* design. Since each factor has two levels, this is a 2^2 design requiring four unique *treatments* or *runs*. (The base 2 refers to the number of levels; the exponent refers to the number of factors.) These four runs are the four combinations of the levels of the factors using either the labels *high* and *low* or plus and minus signs to identify the actual levels:

1. A low; B low = $A_- B_-$
2. A low; B high = $A_- B_+$
3. A high; B low = $A_+ B_-$
4. A high; B high = $A_+ B_+$

A *run* or *treatment* is the unique combination of the factor levels. Note that each run may be performed more than once. The procedure of performing more than one trial of each run is referred to as *replication* when each experimental trial utilizes a completely new setup. A *replicate* is an independent and random application of the run, including the setup. This is considerably different from a *repeat*, which is a repetition of a run without going through a new setup. The 2^2 design defines four treatments or runs. To reduce the impact of the inherent variation in the process, each run is replicated for a total of eight trials. These eight trials must be carried out in a random order to minimize the risk of bias in the results due to unknown or uncontrolled factors.

Randomization refers to the order in which the trials of an experiment are performed. Randomization can be achieved by numbering the trials and then drawing numbers from a hat, using a table of random numbers, shuffling numbered cards, etc. This is important to protect against uncontrolled and/or unknown influences of variables that are not part of the experiment. As an example, assume that an experiment has a single factor, pressure. Assume also that, unknown to the experimenters, a thermostat reading is drifting steadily downward and that the background temperature affects the response. If all the low-pressure trials are performed on day one and the high-pressure trials on day two, is their difference due to the change in pressure or to the change in temperature? The experimenters cannot be sure and, in fact, may not even know there is a problem. If the temperature impact is major, one could erroneously conclude that pressure is a causative factor. In order to minimize this risk of unknown influence, the experimenters randomly assign the order of testing to improve the chances of averaging out this *bias* or *distortion* of the responses related to the factor(s) under study.

There are times one cannot randomize due to physical or cost constraints. Such cases generally lead to *blocking* of the experiment into sections defined by the factor that cannot be randomized. For instance, there may not be enough material from one batch of raw material to complete the experiment. Instead, one may *block* by carrying out a carefully defined set of trials with one batch and the remaining trials with a second batch. Such blocking minimizes the risk of the nuisance-factor batches creating excessive estimates of the inherent variation.

This first part of Chapter 2 is the *design* part of DOE: factor selection, setting levels, defining treatments, randomization of the order of performance. Now the eight trials must be performed exactly as required, keeping all other factors constant. The responses or results that were obtained are shown in Table 2.1.

TABLE 2.1. Data for Example 1.

		A. Polymer Temperature, °F	
		580 (Low) −	600 (High) +
B. Vendors	Vendor Y (Low) −	18.6 17.4 $\overline{Y} = 18.00$	17.5 16.5 $\overline{Y} = 17.00$
	Vendor X (High) +	18.2 16.7 $\overline{Y} = 17.45$	22.9 22.2 $\overline{Y} = 22.55$

Notice the use of plus and minus signs as another way to indicate the two levels. Analysis of these data will be done using the eight-step analytical procedure that follows. These eight steps will work for experiments with any number of factors, as long as the factors have two levels. At first glance, the limitation to two levels may seem very restrictive. In actuality, it permits a remarkable efficiency in the number of trials needed. The primary penalty is that the relationship between a response and a factor with two levels is assumed to be linear. Techniques for identifying nonlinear relationships (curvature) are presented later.

The Eight Steps for Analysis of Effects

1. Calculate effects.
2. Make a Pareto chart of effects.
3. Calculate the standard deviation of the experiment, S_e.
4. Calculate the standard deviation of the effects, S_{eff}.
5. Determine the t-statistic.
6. Calculate the decision limits and determine the significant effects.
7. Graph significant effects.
8. Model the significant effects.

The Analytical Procedure

1. Calculate Effects.

Main effects are defined as the difference in the average response between the high and low levels of a factor. The *Effect of A* is written as *E(A)*. Using plus and minus signs to represent high and low levels of a factor, main effects are defined as

$$E(A) = \overline{Y}_{A+} - \overline{Y}_{A-} = \frac{22.55 + 17.00}{2} - \frac{17.45 + 18.00}{2} = 19.78 - 17.73 = 2.05$$

The main effect of +2.05 means that the average bond strength at the high level of temperature (600°F) was 2.05 units higher than the average bond strength at the low temperature (580°F). The same data can now be used to determine the main effect of factor B (vendor) on the process. Note that a plus sign is used to designate vendor X while a minus sign is used to designate vendor Y. This is an arbitrary choice that once chosen must be used consistently.

$$E(B) = \overline{Y}_{B+} - \overline{Y}_{B-} = \frac{22.55 + 17.45}{2} - \frac{17.00 + 18.00}{2} = 20.00 - 17.50 = 2.50$$

This indicates that the paper supplied by vendor X averaged 2.50 units higher in bond strength than that supplied by vendor Y. *An effect is the difference in averages.* Note that in a DOE the same data are used for more than one factor. This is the fundamental concept of DOE: an experiment is used to define many effects, rather than performing an experiment for each factor. The graphs in Figures 2.1 and 2.2 indicate the direction of influence for the factors and communicate the meaning of the effects very clearly.

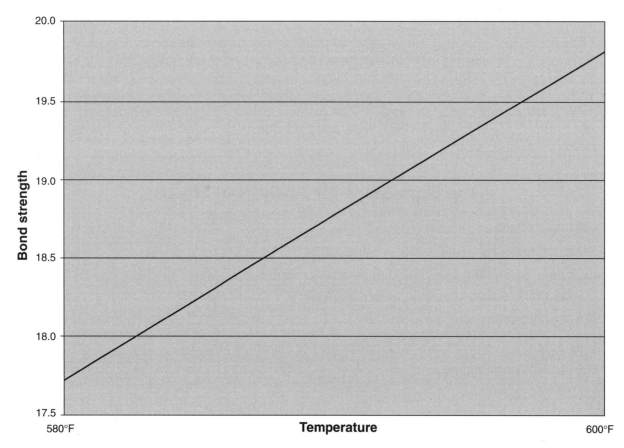

FIGURE 2.1. Effect of temperature on bond strength.

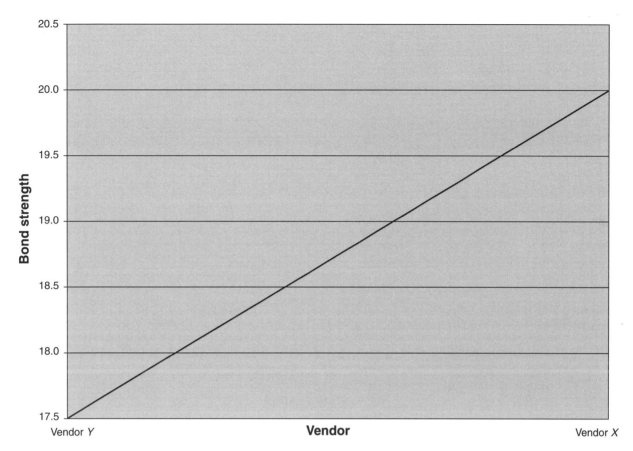

FIGURE 2.2. Effect of vendor on bond strength.

The plots of the main effects of temperature and vendor indicate that a higher temperature results in a stronger bond and that vendor X is superior to vendor Y. Observe the data in Table 2.1. Are those results *always* true and to the amount expected? Are there certain combinations of temperature and vendor that don't seem to provide the expected results? An *interaction* occurs when a particular combination of two factors does something unexpected from simply observing their main effects. An interaction is defined as one-half of the difference between the effect of A at the high level of B and the effect of A at the low level of B. Mathematically, this is

$$E(AB) = \tfrac{1}{2}[(\overline{Y}_{A+} - \overline{Y}_{A-})_{B+} - (\overline{Y}_{A+} - \overline{Y}_{A-})_{B-}]$$

Using the data in the table,

$$E(AB) = \tfrac{1}{2}[(22.55 - 17.45) - (17.0 - 18.0)] = \tfrac{1}{2}[5.1 - (-1.0)] = 3.05$$

The interaction effect essentially increases or decreases the main effect in the experiment by 3.05 units of strength. For example, the main effect of temperature equals +2.05. However, when paper from vendor X is used, the effect of temperature is actually 5.10; i.e., (22.55 − 17.45). Conversely, when paper from vendor Y is used, the temperature effect is −1.0; i.e., (17.0 − 18.0). This means that when using paper from vendor Y, higher bond strengths are achieved with *lower* temperature. The plot reveals what the interaction effect means (see Figures 2.3 and 2.4 on page 10).

Interactions can be plotted as *vendor-temperature* or as *temperature-vendor*; both are correct and are simply two views of the same phenomenon. The two plots do *not* always look alike. If in doubt about which is more useful, plot both!

A final precaution on interactions: if an interaction proves to be significant, the interaction chart is more important than the main effect charts. If there is an interaction, the main effect describes average results whereas the interaction is more appropriate in describing the joint effect.

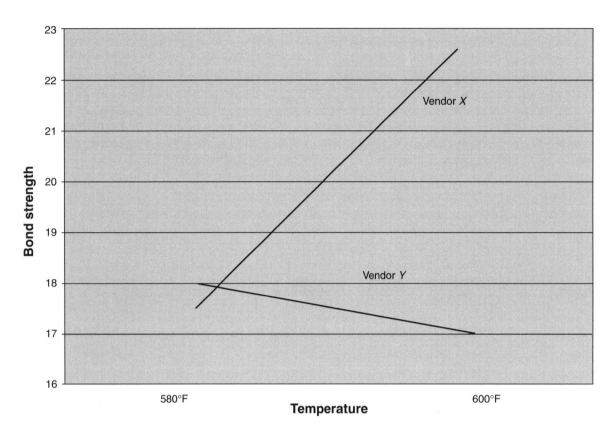

FIGURE 2.3. Temperature-vendor interaction.

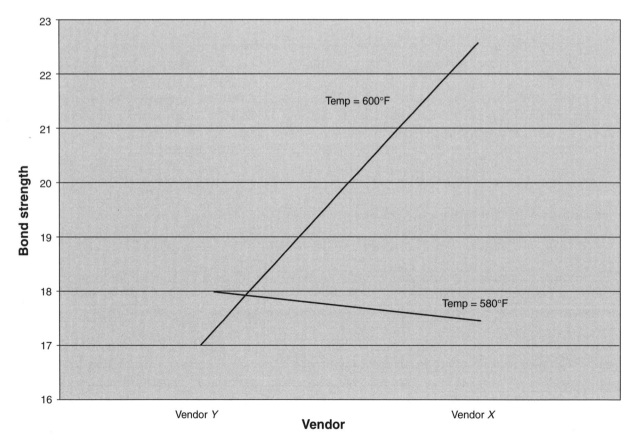

FIGURE 2.4. Vendor-temperature interaction.

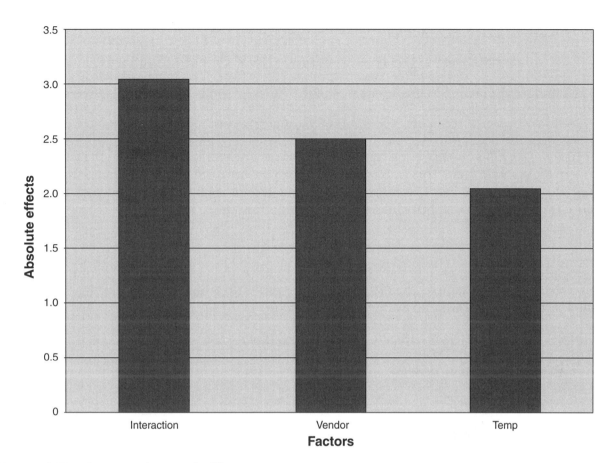

FIGURE 2.5. Pareto chart of effects.

2. Make a Pareto Chart of Effects.

To show the relative importance of the effects, plot their absolute value as a Pareto chart (see Figure 2.5). The Pareto chart, which becomes more useful when larger experiments are studied, is often sufficient to determine what effects are meaningful.

The problem now is to test these effects for significance; i.e., how do we show that the results of the experiment are beyond the inherent variation to be expected in the experiment? The next four steps will determine whether these three effects are significant or meaningful.

3. Calculate the Standard Deviation of the Experiment, S_e.

Determination of significance requires calculation of a *standard deviation* as a measure of the *inherent variation* or *experimental error* in the process. In SPC, this was referred to as *sigma* and used the symbol σ_e. In DOE, the symbol S_e is used. To obtain S_e, the *variance* (S^2) is calculated for each run or treatment. (Here is where a scientific calculator will be worth its cost! Just be sure to use the key for a divisor of $n-1$.) These variances are then averaged and converted to a standard deviation (S_e) by taking the square root.

Let's have a tutorial on this. A *variance* is the square of the deviation of each observation of a sample from the sample average. The formula is

$$S^2 = \frac{\Sigma(X_i - \overline{X})^2}{n-1}$$

Assume three responses: 12, 14, 16. Since the average is 14,

$$S^2 = \{(12-14)^2 + (14-14)^2 + (16-14)^2\}/(3-1) = \{(-2)^2 + 0^2 + 2^2)\}/2 = 4.0$$

The standard deviation S is the square root of the variance and is a measure of inherent variation.

$$S = \sqrt{S^2} = \sqrt{4} = 2.0.$$

Don't plan on doing this by hand; use a calculator! Variances are averaged even though standard deviations are the statistic used. This is to permit creation of a measure of spread or variation that will always be a positive number. The variances are calculated for each run and averaged. The average variance is converted to a standard deviation by taking the square root:

$$S_e = \sqrt{(\Sigma S_i^2 / k)}$$

where S_i^2 is calculated for each of the k runs. In this exercise, the variances of the four runs of Table 2.1 are calculated. For example, the two trials for the run with temperature = 580° F and vendor Y resulted in 18.6 and 17.4 with an average of 18.0. Then

$$S_i^2 = [(18.6 - 18.0)^2 + (17.4 - 18.0)^2]/(2-1) = (.36 + .36)/1 = .72$$

The variances are shown in Table 2.2.

TABLE 2.2. Variances for each run of example 1.

	Temperature	
	580°F	600°F
Vendor Y	.720	.500
Vendor X	1.125	.245

$$S_e^2 = (.72 + .50 + 1.125 + .245)/4 = 0.648$$

and

$$S_e = \sqrt{.648} = .80$$

This estimate of the inherent variation represents (1) measurement variation, plus (2) the inability of the experimenter to repeat the conditions, plus (3) the inability of the process to repeat the same response for the same conditions.

There is more than one way to calculate the standard deviation of the experiment. This technique is used since it is the classic approach and more closely matches what will later be seen in software. Range techniques from SPC could also be used with little loss in precision of the estimate, that is,

$$S_e = \frac{\overline{R}}{d_2}$$

where $\overline{R}$ is the average of the ranges of the runs and d_2 is a tabular value based on the number of replicates.

4. Calculate the Standard Deviation of the Effects, S_{eff}.

Effects are differences between averages, requiring a modified standard deviation. This is called the *standard deviation of the effects* and is defined as

$$S_{eff} = S_e \sqrt{(4/N)}$$

where N is the total number of trials. This formula will hold for any number of trials as long as the factors have two levels. In the example,

$$S_{eff} = .80\sqrt{(4/8)} = 0.57$$

5. Determine the t-Statistic.

To use the t-table, the *degrees of freedom* in the experiment must be determined. Degrees of freedom (d.f.) measure the amount of information available to estimate the standard deviation. The calculation is

$$d.f. = (\text{\# of observations per run} - 1) \times (\text{\# of runs}) = (2-1) \times (4) = 4$$

Next, a reference number must be selected from the t-table on page 66 in the appendix. Ninety-five percent confidence is customarily considered necessary to claim significance in effects. Confidence is

$$1 - alpha\ risk$$

where the alpha risk is the chance of erroneously claiming significance. The typical alpha risk (also written with the Greek symbol α) is therefore 5 percent. This table is based on that level of confidence. Referring to the t-table for 4 degrees of freedom and α = 5 percent, the tabular value is $t = 2.78$.

6. Calculate the Decision Limits and Determine the Effects.

The *decision limits (DL)* for the significance of effects in this DOE can now be calculated. The question is whether the computed effects are significantly different from zero and, therefore, not due to random variation. If the effects are outside the zone defined by the decision limits, the effects are considered real or significant. The decision limits are calculated by

$$DL = \pm (t_{\alpha,df})(\sigma_{eff}) = \pm (2.78)(.57) = \pm 1.58$$

The results are shown graphically in Figure 2.6.

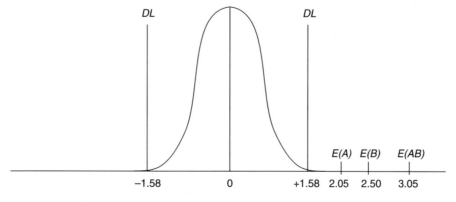

FIGURE 2.6. Decision limits.

What is the conclusion here? All three of the effects in the example exceed the limits and are judged to be real or significant. Since the interaction effect was significant, one must jointly consider the two factors to determine how to optimize the process. Since the objective was to maximize bond strength, one should use paper from vendor X and operate at a temperature of 600°F. Refer to Figures 2.3 and 2.4.

7. Graph Significant Effects.

Actually, graphs need only be made for significant effects. If an effect is not significant, the graph is of no interest. If an interaction is significant, that graph is more meaningful than those of the individual factors since an interaction means that the experimenter must consider the factors jointly. The graphs for this example have already been shown in Figures 2.1, 2.2, 2.3, and 2.4.

8. Model the Significant Effects.

The *model* or *prediction equation* is useful to predict the optimum outcome for future validation experiments. The model is a linear equation of the following form, using only the significant effects. However, if an interaction effect is significant, the terms for the two main effects are also included *even if they are not significant*. This is due to the *hierarchy rule* for defining a model. This rule becomes more important in advanced techniques for optimization. The levels of A, B, and AB are coded by -1 for the low levels and $+1$ for the high levels. The term $\bar{\bar{Y}}$ represents the average of all the data.

$$\hat{Y} = \bar{\bar{Y}} + \frac{E(A)}{2}A + \frac{E(B)}{2}B + \frac{E(AB)}{2}AB$$

Since our objective is to maximize bond strength, A and B are set at the high (+) levels that will provide maximum strength. Specifically,

$$\hat{Y} = 18.75 + (2.05/2)(+1) + (2.50/2)(+1) + (3.05/2)(+1)(+1) = 22.55$$

If the best combination of temperature and vendor paper is used, bond strength is estimated to average 22.55. This is better than the results predicted simply from the main effects of A and B due to the contribution of the interaction term. An interaction either increases the response over that expected due to main effects alone or it reduces the response below that expected from main effects.

What if one needed to know the expected response for a temperature of 595°F with paper from vendor X? The model uses coded values for the levels so the temperature of 595°F must be changed to coded values in order to interpolate the results:

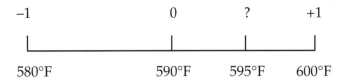

Since 1 unit = 10° and 595° is 5 degrees from the zero point equivalent to 590°, the coded equivalent of 595° is 5/10 = + .5. This value is then inserted into the model for A with $B = +1$:

$$\hat{Y} = \bar{\bar{Y}} + \frac{E(A)}{2}A + \frac{E(B)}{2}B + \frac{E(AB)}{2}AB$$

$$= 18.75 + (2.05/2)A + (2.50/2)B + (3.05/2)AB$$

$$= 18.75 + 1.03A + 1.25B + 1.53AB$$

$$= 18.75 + 1.03(.5) + 1.25(1) + 1.53(.5)(1) = 21.28$$

This interpolation procedure is useful when there are only a couple of significant factors. Otherwise, too many assumptions must be made since one would have a single equation in several unknowns.

What further study can be recommended if even higher bond strength is desired? Does the temperature effect continue beyond 600°F? It is dangerous to extrapolate beyond the range of study (580° to 600°F). Plan a verification study at 600°F and at 610°F with paper from vendor X. This will verify the previous conclusions and explore the potential for even better performance.

The objective in this example was to achieve stronger bond strength. Suppose the objective had been different. What if the bond strength problem was not one of inadequate strength but of too much variation in the strength since temperature could not be tightly controlled? What conclusions would be drawn in that circumstance? What conditions would make the process *robust* or insensitive to changes in the process variables of temperature and vendor? The interaction plots show that changes in temperature have little effect on bond strength if paper from vendor Y is used. This is why the objective of the study must be clearly understood before the experiment is finalized.

For demonstration purposes, again assume a different objective. Suppose that, instead of achieving a maximum or minimum, the experimenter needed to hit a target response of 20. This would lead one to determine the setting of the quantitative (numeric) factor of temperature. Based on the interaction graph (Figure 2.4), note that only vendor X material provides a range of responses that includes the value of 20. If B is set at +1 for vendor X, what setting for factor A (temperature) would provide a forecast of the target value? To determine this, solve the model for A where $B = +1$ (vendor X) and $Y = 20$ (target response):

$$\hat{Y} = \bar{\bar{Y}} + \frac{E(A)}{2}A + \frac{E(B)}{2}B + \frac{E(AB)}{2}AB$$

$$20 = 18.75 + 1.03A + 1.25B + 1.53\,AB$$

$$20 = 18.75 + 1.03A + 1.25 + 1.53\,A$$

$$2.56A = 0$$

$A = 0$ in coded units or, by interpolation, $A = 590°$ in actual units.

This completes the analytical procedure. It is very important that the conclusions from this experiment be verified either by operating at the recommended conditions or by further experimentation. (More advanced techniques are covered in the references on page 63 to more completely address the subject of optimization.) All reports, graphs, and recommendations should be put in layman's terms, not in the statistical terminology that is used in the analysis. Remember that the average employer may not know what A– means, but he or she understands Pareto charts (the 80/20 rule) and graphs as long as the references are in familiar language. A box with the steps of the analytical procedure is provided for quick reference.

THE EIGHT STEPS FOR ANALYSIS OF EFFECTS

1. Calculate effects.
2. Make a Pareto chart of effects.
3. Calculate the standard deviation of the experiment, S_e.
4. Calculate the standard deviation of the effects, S_{eff}.
5. Determine the t-statistic.
6. Calculate the decision limits and determine the effects.
7. Graph significant effects.
8. Model the significant effects.

These steps are the fundamental analysis techniques for any number of factors at two levels. Remember that small DOEs such as the previous example are usually used in refining experiments after larger screening experiments or when some quick troubleshooting is needed. The spacing of the quantitative levels (temperature in our example) is also important. In a screening design, the levels should be set as wide or as bold as practical to make it easier to discover significant factors for later study. In a refining design, the levels would be set more closely and more replication would be required since the size of the effects would be reduced. Additional data would be needed to demonstrate significance of the smaller effects.

A final caution is needed on the statistical control of the process. Lack of control increases the experimental error and can also create false effects. It is important to stabilize the process, i.e., eliminate the special causes. Otherwise, replication must be increased to overcome the distortion (overestimate) of the estimate of inherent variation (S_e) due to the presence of special causes of variation.

Review of the Experimental Procedure

Having completed an experiment and gone through the analytical procedures, one can review the broader concepts of experimentation. There are a series of *logic steps* that must be addressed as one prepares to launch a DOE.

1. What is the process to be studied? How broadly or narrowly is it defined? A flowchart is a good tool for this analysis.

2. What is the response? What needs to be improved? Should there be more than one response? Note that additional responses are free, costing only the measurements!

3. What is the measurement precision? Is there bias in the measurement system? Has a measurement analysis been completed? Is it adequate?

4. Generate candidate factors. This is best done with a small team using brainstorming after a review of all available data and information on the process and response variables. A cause-and-effect diagram with a flowchart of the process is useful with this brainstorming. The trick is to be innovative, to think outside usual boundaries, and yet not try to reinvent proven technology. Provide opportunities for surprises! The team should be knowledgeable about the issues and follow the rules for brainstorming.

5. Determine the levels for the factors selected for the DOE. In screening experiments, the rule is to have levels broadly spaced but not to the point of being foolhardy. In refining experiments, levels will be much tighter and will require more replication.

6. Select the experimental design. This is the set of treatments or runs that will be performed. This also includes deciding on the amount of replication. Finally, the randomized order of the trials is determined. (*Randomization* is the insurance policy against misleading conclusions due to outside influence during the experiment.)

7. Establish a plan to control (or at least monitor) extraneous variables.

8. Perform the experiment according to the design. The DOE *must* be carried out per its design. Identify trial materials carefully. Keep good notes.

9. Analyze, draw conclusions, and assess process impact. What process variables can be changed—and how—to improve the process?

10. Verify and document the new process as defined by the experiment.

11. Propose the next study for continuation of this project, or declare the project complete. Make sure that all reports that go beyond the team are in language and terminology that are easily understood.

Example 2: Water Absorption in Paper Stock

One of the critical measures of some raw paper is its ability to minimize the absorption of water. This is tested by placing a carefully weighed sample under water for a specific time and then reweighing the sample. A percentage weight gain is the measure of the water absorption characteristic. In an effort to minimize this absorption, two candidate factors have been selected for experimentation:

1. Factor A is the type of *sizing* chemical used. (Sizing is an expensive chemical inhibitor of water absorption.) Two different types of chemicals are to be evaluated. These are *normal*—been in use for ages—and *enhanced*—a new formulation.
2. Factor B is the amount of the sizing chemical to add to the slurry. Typically, 5 percent is added. It has always been assumed that adding more will improve (reduce) the water absorption accordingly. The trial will compare 5 percent and 8 percent as additive rates.

The objective is to get water absorption below 7 percent. The design is a 2^2 factorial (two levels and two factors) with four replicates. The design contains four unique runs or treatments and 16 total trials. The results in percent water absorption are shown in Table 2.3.

TABLE 2.3 Data for Example 2.

		(A) Sizing Type	
		Normal (−)	Enhanced (+)
(B) Amount of Size	5% (−)	1.34 5.24 6.56 4.20 $\overline{Y} = 4.34$ $S_1^2 = 4.92$	10.38 6.66 6.68 9.56 $\overline{Y} = 8.32$ $S_2^2 = 3.74$
	8% (+)	4.98 3.49 3.92 5.29 $\overline{Y} = 4.42$ $S_3^2 = .73$	6.22 11.0 4.72 8.36 $\overline{Y} = 7.58$ $S_4^2 = 7.45$

The analytical procedure must now be applied. Verify these results using the eight-step analytical procedure explained earlier.

1. Calculate Effects.

Main effects are indicated by the following equations.

$$E(A) = \overline{Y}_{A+} - \overline{Y}_{A-} = (8.32 + 7.58)/2 - (4.34 + 4.42)/2 = 7.95 - 4.38 = +3.57$$
$$E(B) = \overline{Y}_{B+} - \overline{Y}_{B-} = (7.58+4.42)/2 - (8.32+4.34)/2 = 6.00 - 6.33 = -.33$$

The graphs in Figures 2.7 and 2.8 illustrate the meaning of the effects.

The interaction effect is illustrated in Figure 2.9 and is determined by the equation

$$E(AB) = \tfrac{1}{2}[(\overline{Y}_{A+} - \overline{Y}_{A-})_{B+} - (\overline{Y}_{A+} - \overline{Y}_{A-})_{B-}]$$
$$= \tfrac{1}{2}[(7.58 - 4.42) - (8.32 - 4.34)] = \tfrac{1}{2}[-.82] = -.41$$

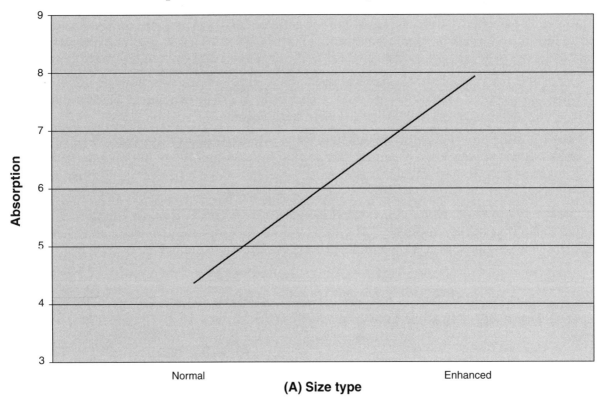

FIGURE 2.7. Effect of size type.

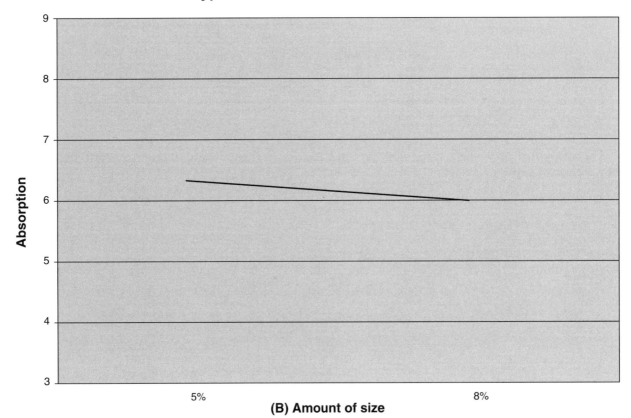

FIGURE 2.8. Effect of amount of size.

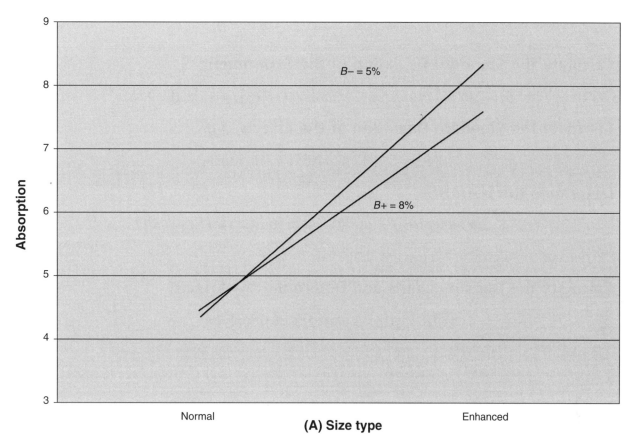

FIGURE 2.9. Interaction of type and amount.

2. Make a Pareto Chart of Effects

Could one draw conclusions at this point if necessary? Observe that a reasonable decision could be made simply by using the Pareto principle (see Figure 2.10).

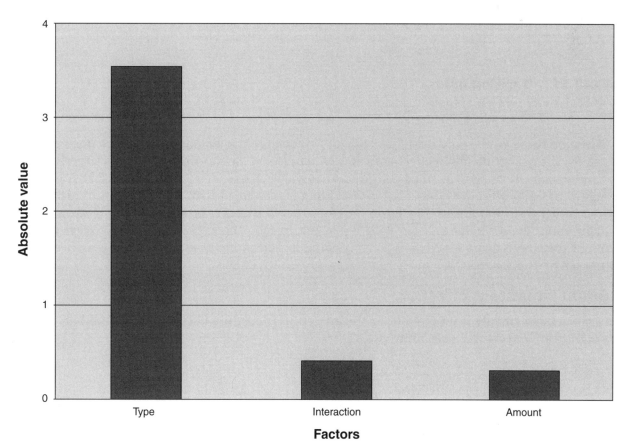

FIGURE 2.10. Pareto of effects.

3. Calculate the Standard Deviation of the Experiment, S_e.

$$S_e = \sqrt{(\Sigma S_i^2 / k)} = \sqrt{[4.92 + 3.74 + .73 + 7.45]/4} = \sqrt{4.21} = 2.05$$

4. Calculate the Standard Deviation of the Effects, S_{eff}.

$$S_{eff} = S_e \sqrt{4/N} = 2.05\sqrt{4/16} = 1.03$$

5. Determine the t-Statistic.

$$df = (\text{\# replicates per run} -1) \times (\text{\# of runs}) = (4-1) \times (4) = 12$$

For 95 percent confidence or a 5 percent alpha risk, $t = 2.18$.

6. Calculate the Decision Limits and Determine the Effects.

$$DL = \pm(t)(S_{eff}) = \pm(2.18)(1.03) = \pm 2.25.$$

Graphically, any effect outside the decision limits is significant. Refer to Figure 2.11.

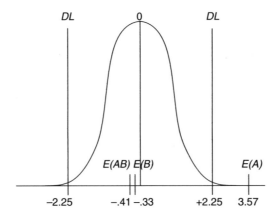

FIGURE 2.11. Decision limits.

Based on these limits, only the type of sizing will change the water absorption characteristics of the paper. The plot of the main effect indicates which is best: the normal additive is superior to the new and enhanced version! Further, consider what was learned about the amount of sizing to add. No significant difference was found between adding 5 percent and adding 8 percent. Which would one recommend since it doesn't matter to the response? If this is expensive material, why add cost by using more than is essential? Go with 5 percent. In fact, the experimenter may wish to run a refining experiment using less than 5 percent sizing to determine how low it can be reduced without impact on the absorption.

7. Graph Significant Effects.

The graphs have already been plotted. As the analytical procedure becomes more familiar, plotting only significant effects will save time.

8. Model the Significant Effects.

The model is very simple:

$$\hat{\hat{Y}} = \overline{\overline{Y}} + \frac{E(A)}{2}A = 6.17 + (3.57/2)A$$

where A is -1 for normal and $+1$ for enhanced sizing. Using normal material ($A = -1$), an average performance of 4.39 is predicted. Since the objective was to stay below 7.0 percent, how well will this process perform if the average is 4.39? This is a capability issue. Recall from basic SPC that a measure of process capability is C_{pu}, which is calculated by

$$C_{pu} = (\text{upper spec limit} - \overline{\overline{Y}})/(3\ S_e) = (7.0 - 4.39)/(3 \times 2.05)$$

$$= 2.61/6.15 = .42.$$

Since any capability index should be > 1.33, this value of C_{pu} indicates an undesirable situation even after identifying the best operating condition from the experiment. This is illustrated graphically in Figure 2.12.

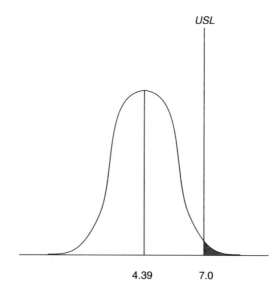

FIGURE 2.12. Process capability.

The interest in trying other products is understandable. This is a poor process capability since a substantial amount of the product will exceed the specification. This situation can only be improved by lowering the average and/or reducing the inherent variation. At this point, the team must go back to the drawing board to try to find other candidates for improving this process since these alternatives to the process would not be beneficial.

A Different Analytical Technique: The Spreadsheet Approach

The approach used in step 1 to calculate effects helps communicate the meaning of the terms *main effects* and *interaction effects*. As the number of factors increases, that procedure becomes very awkward.

An easier technique is presented here for use in the future. This technique uses the notation of a minus sign (–) and a plus sign (+) to indicate the low and high levels of each factor. Consider again the data for the last example as it is presented in Table 2.4.

TABLE 2.4. Restatement of data from Example 2.

		(A) Size Type	
		Normal (–)	Enhanced (+)
(B) Amount of Chemical	5% (–)	(– –) 1.34 5.24 6.56 4.20 $\bar{Y} = 4.34$ $S_e^2 = 4.92$	(+ –) 10.38 6.66 6.68 9.56 $\bar{Y} = 8.32$ $S_e^2 = 3.74$
	8% (+)	(– +) 4.98 3.49 3.92 5.29 $\bar{Y} = 4.42$ $S_e^2 = .73$	(+ +) 6.22 11.0 4.72 8.36 $\bar{Y} = 7.58$ $S_e^2 = 7.45$

Each run can be described by the combination of signs representing the levels of the factors for that treatment. The order of the signs is always alphabetical; i.e., – + represents an *AB* combination with *A* at the low level and *B* at the high level. The data can be regrouped as shown in Table 2.5 with the averages and variances entered for each run.

TABLE 2.5. Example 2 in spreadsheet format.

Treatments	A	B	AB	$\bar{Y}$	S^2
1	–	–	+	4.34	4.93
2	+	–	–	8.32	3.72
3	–	+	–	4.42	.72
4	+	+	+	7.58	7.45
ΣY_+	15.90	12.00	11.92		
ΣY_-	8.76	12.66	12.74		$S_e = 2.05$
$\bar{Y}_+$	7.95	6.00	5.96		$\bar{Y}_+ = \Sigma Y_+ /$(half the no. of runs)
$\bar{Y}_-$	4.38	6.33	6.37		$\bar{Y}_- = \Sigma Y_- /$(half the no. of runs)
Effect	+3.57	–0.33	–0.41		Effects = $\bar{Y}_+ - \bar{Y}_-$

The steps for this procedure are:

1. Each row of the interaction column (*AB*) has the sign of the product of the signs of the two interacting factors (*A* and *B*) in that row.

2. For each column, the average response is summed for each row containing a plus sign in that column and is entered in the ΣY_+ row.

3. For each column, the average response is summed for each row containing a minus sign in that column and is entered in the ΣY_- row.

4. For each column, the $\overline{Y}_+$ and the $\overline{Y}_-$ are obtained by dividing the ΣY_+ and the ΣY_- by half the number of runs or treatments.
5. For each column, the effect is calculated by $\overline{Y}_+ - \overline{Y}_-$.

Note that the same answer was obtained here as in the original calculations. The first approach makes the meanings more obvious but is difficult to manage as the number of factors increases. The spreadsheet approach is faster and can be expanded to any number of factors very easily. It is easily handled by a spreadsheet. This spreadsheet procedure will be used from now on. To plot significant main effects from this table, plot the $\overline{Y}_+$ and the $\overline{Y}_-$ for a factor. However, it will still be necessary to make a 2 × 2 table of averages for each interaction in order to plot the interactions.

Exercise 1: Coal-Processing Yield

In a coal-processing plant, pulverized coal is fed into a flotation cell. There the coal-enriched particles are attached to a frothing reagent and floated over into the next wash cell. Rock or low-grade coal particles sink and are taken out the bottom as waste. The yield is the proportion of the input coal that is floated into the next washing step (see Table 2.6).

TABLE 2.6. Data for coal yield.

		(A) Retention Time	
		45 sec. (−)	60 sec. (+)
(B) Flotation Reagent Concentration	1% (−)	(− −) 57.7 61.5 57.9 53.5 $\overline{Y} = 57.65$ $S^2 = 10.70$	(+ −) 64.5 71.8 69.1 73.4 $\overline{Y} = 69.7$ $S^2 = 15.16$
	2% (+)	(− +) 80.1 83.1 85.1 77.9 $\overline{Y} = 81.55$ $S^2 = 10.14$	(+ +) 79.5 70.2 77.2 73.5 $\overline{Y} = 75.10$ $S^2 = 16.78$

In attempting to enrich the yield of the cell, two factors are being considered:

1. Retention time in the cell: 45 seconds versus 60 seconds
2. Flotation reagent concentration in the solution: 1 percent versus 2 percent

A 2^2 factorial design was used with four replicates of each of the four runs. The objective is to maximize the yield in this processing step.

Analyze this experiment using the eight-step procedure for analysis of effects. Use the spreadsheet approach. State your conclusions and recommendations. After you have completed your work, refer to the results in the appendix.

Nonlinear Models

Restricting the experiments to two levels allows simplification of the analyses and provides substantial reduction in the number of runs required in larger experiments. However, this permits only linear (straight-line) relationships to be defined. This is generally sufficient for most experimental work until the final refining or optimization stages of a study. At that time, three or more levels may be used to identify a *nonlinear* or quadratic relationship. *Center points* can also be used to detect the presence of nonlinearity or curvature. These are trials run at the midpoint of the factor levels—the zero level. (Obviously, nonlinearity only has meaning with quantitative factors.) Center points also provide an opportunity to increase the degrees of freedom and improve the estimate of the experimental error (S_e). The following procedures apply in addressing nonlinearity:

1. *Modify the procedure for estimating the experimental error.*

    a. Calculate and average the variances of the base (non–center point) runs as usual. Then calculate the variance of the center points. Define these variances as S_b^2 and S_c^2 respectively.

    b. Calculate the degrees of freedom for the base design and for the center points. Degrees of freedom for the base design = df_b = (number of runs) × (number of replicates − 1). Degrees of freedom for the center points = df_c = (number of center points −1).

    c. Calculate S_e^2 as the weighted average of the two variance estimates, weighting each by its degrees of freedom.

    $$S_e^2 = [(df_b \times S_b^2) + (df_a \times S_c^2)]/(df_b + df_c)$$

    $$S_e = \sqrt{S_e^2}$$

2. *Define the nonlinear effect.*

    $$E(\text{nonlinearity}) = \overline{\overline{Y}} - \overline{Y}_{\text{center}}$$

    where $\overline{\overline{Y}}$ is the average of all trials except center points, and $\overline{Y}_{\text{center}}$ is the average of the center points. If the relationships are linear, the center points should produce an average response that approximates the grand average ($\overline{\overline{Y}}$). If the average of the responses at the center point is significantly different from the grand average, nonlinearity exists.

3. *Define the standard deviation to test nonlinearity.*

    $$S_{\text{nonlin}} = S_e \sqrt{1/N + 1/C}$$

    where N is the total number of trials *not at the center* and C is the number of trials at the *center*.

4. *Determine the t-statistic using all the degrees of freedom.*

    $$\text{Degrees of freedom} = df_b + df_c$$

    This was the reason for adding the additional calculations using center points. The extra degrees of freedom will provide a stronger t-statistic. *This t-statistic can be used for all effects, nonlinear and basic.*

5. *Calculate a decision limit for the nonlinear effect.*

    $$DL = \pm t \times S_{\text{nonlin}}$$

The following guidelines must be remembered when dealing with nonlinearity:

1. The presence of center points changes the calculation of S_e, degrees of freedom, and the t-statistic for all assessments. The increased complexity is the price of increasing the degrees of freedom. Since the expense of an experiment usually leads to minimum degrees of freedom, the additional complexity is worth it.

2. If the nonlinearity effect is significant, the linear plots for main effects are suspect. The test for nonlinearity does *not* tell the experimenter which factor(s) contain the curvature, only that it exists. Significance of main effects is no longer reliable.

3. Interpolation of the linear model is no longer wise.

4. Significant nonlinearity must be evaluated by further experimentation: (a) expanding the current experiment to three levels or (b) repeating the DOE from the beginning with three levels.

5. There should be at least four center points to have a moderate chance of detecting nonlinearity.

Exercise 2: Corrosion Study

Analyze this exercise for all effects, including nonlinearity. State your conclusions and recommendations for further experimentation. Use the spreadsheet procedure of Example 2 and the procedures and guidelines for nonlinearity. After you have completed your work, refer to the results in the appendix on page 72.

A research group has been charged with developing a steel with better corrosion resistance than the current product. Previous experimentation has established the chemical composition of all but two elements. Chromium and nickel levels have been narrowed but not finalized. The team wishes to evaluate: (1) chrome with levels of .04 percent and .10 percent, and (2) nickel with levels of .10 percent and .20 percent. The design was a 2^2 factorial with three replicates. Since there was a concern over nonlinearity, five center points were run at the factor settings of Cr at .07 percent and Ni at .15 percent. The order of the seventeen experimental trials was randomized, and the response is the weight loss in an acid bath during a fixed time period. The objective is to minimize the weight loss. Data are given in Table 2.7. Have fun!

TABLE 2.7. Data for weight loss by corrosion.

		(A) Chromium	
		.04% (−)	.10% (+)
	.10% (−)	4.1	5.5
		4.9	4.0
(B) Nickel		4.3	4.3
	.20% (+)	9.6	8.2
		8.7	7.4
		10.1	8.7

Responses at center point: 9.5, 9.0, 8.6, 8.7, 8.1

Process Quality Improvement

A Systematic Approach

> The issues of organization, planning, action, and continuous process improvement are the keys to successful application of the tools at hand.

By Praveen Gupta, Quality Technology Co., Carol Stream, Illinois

Quality, not volume, has become the number one measure of customer satisfaction in the manufacturing and service sectors. Though accounting systems can determine the price of quality in terms of manufacturing cost, they exclude business lost because of poor quality.

The necessary pace of quality improvement is reflected by ever-changing customer requirements. Today's "luxuries" become tomorrow's necessities. In order to fully satisfy the customer, the product must be developed according to his requirements. However, it is equally important to develop a product that is manufacturable.

Inefficiencies in manufacturing result in productivity and quality problems. The greatest designs can become manufacturing disasters. Poor product quality can be traced to poor designs, limited process capability, poor preventive maintenance, variations in the incoming material, and untrained employees. In a nutshell, quality problems are caused by a lack of commitment to be the best in every aspect of the business.

The three aspects of solving a quality problem are containment, correction, and prevention. Containment means customer appeasement. This process ensures that the customer never sees the problem. Correction is like firefighting, because the problem is an emergency to be solved as soon as possible. Prevention is long-term problem solving. Most likely the solution results either in a product or process design change. In order to prevent quality problems from recurring, a continuous quality improvement strategy is mandated.

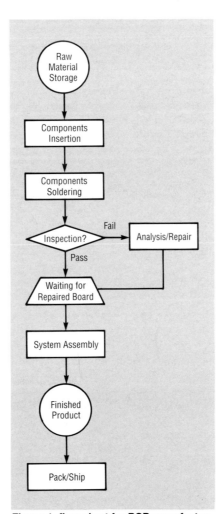

Figure 1, flow chart for PCB manufacturing identifies critical stages that affect product quality.

Improvement Strategy

One of the many roadblocks in implementing a company-wide quality improvement strategy is a perception of high cost. The fact is, continuous quality improvement efforts have resulted in higher profits, productivity, customer satisfaction, and increased market share. This improvement requires a conscious effort to change conventional perceptions of the "cost" of high quality. Company-wide process improvement will occur only with a systematic approach that is supported by management. Key ingredients for improving the quality of product, process, and service are:
- Management commitment, guidance, and participation.
- Short- and long-term improvement plans.
- Cross-functional team and individual involvement.
- A performance assurance system.
- Recognition of success and analysis of failures.

Process Management

An organization must be designed to accomplish the necessary goals. In the past, functional organizations have worked suboptimally with limited success. For optimum performance, organizations are becoming more cross-functional due to the increased complexity of the processes and products. The heirarchy in some companies even has been "inverted," with the CEO at the bottom of the infrastructure bearing the onus of the organization and the customer at the top.

The Malcolm Baldrige National Quality Award has seven categories that address cross-functionally within an entire organization. The categories are leadership, information and analysis, strategic quality planning, human resource utilization, quality assurance of products and services, quality results, and customer satisfaction. In this type of orga-

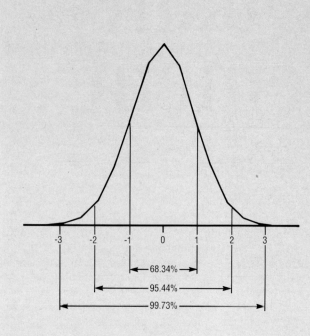

Figure 2, normal distribution curve. Knowing the deviations (±), one can predict variations beyond the specification limits.

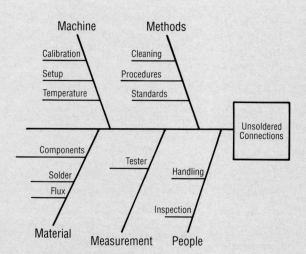

Figure 3, an Ishikawa, or cause-and-effect, diagram is used to isolate reasons for variations (here for soldering defects) and to establish relationships.

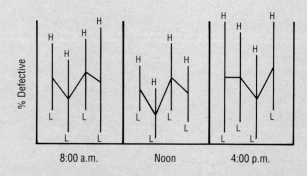

Figure 4, a multi-vary chart traces variations in the processing of a single unit (PCB) during a production day.

nization, the management is in the driver's seat.

The "sixth sense" of management is information. Without information, the ability to make sound decisions is impaired. The data (information) collection process permits management to listen to their processes and people. Lack of data, on the other hand, is the main obstacle for quality improvement efforts in most companies. Companies do not know where they stand or where they want to go. Specific goals must be devised, and a system for data collection and analysis is the first step in improving process quality. To implement a data collection system:

1. State the purpose of collecting data. This should be based on customer requirements or process quality improvement objectives.
2. Develop a data-collection log sheet for collecting time, date, process measurements, and other observations.
3. Identify critical characteristics of the product to be monitored and related process outputs.
4. Map the process and create a flow chart showing storages, waits, and check points (test, inspection, or audit). An example of a PCB manufacturing process flow chart is shown in figure 1.
5. Create data display methods such as trend charts, control charts, histograms, multivariation charts, and cumulative maps.
6. Develop a data-reporting method. This could be either informal communication or a published report.

Variation Analysis

Due to variation in methods, material, machines, or people, the process output will vary over time. The standard deviation is a measure of its variability. In many processes, the output of a high-volume manufacturing process follows a bell-shaped curve, also called a normal distribution curve. The normal distribution curve is characterized by two parameters, the mean and the standard deviation. The area under the normal curve within a specified range around the mean is shown in figure 2. Knowing the standard deviation, one can predict defect levels beyond the specification limits.

In order to prevent defects from occurring in production, the product design should incorporate manufacturing variations. This variability can be related with design specifications using process capability and process performance indices. The process capability, C_p, is the

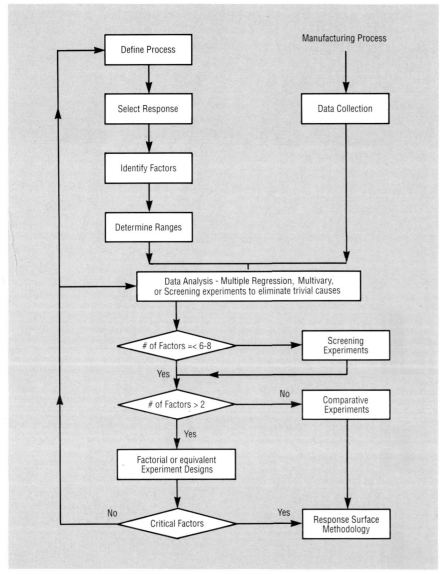

Figure 5, a strategy for running experiments when the number of variables is less than 6.

charts, cumulative charts, histograms, and scatter plots.

The cause-and-effect diagram (figure 3) was developed by Professor Kaoru Ishikawa in the 1950s.[2] The diagram is used to illustrate the various causes affecting a process quality problem. A cause-and-effect diagram is used to sort out the causes of variation and establish mutual relationships.

A multivary chart is used to depict the range of variation within a process. The process variation can be broken down into three categories: within a piece (positional), piece-to-piece (cyclical), and time-to-time (temporal). This partitioning of variation narrows the scope of the problem to a manageable size. Often, this stratification leads to a quick solution. Since using multivary charts does not require special or experimental runs, it is an economic tool for variation analysis. An example of a multivary chart is shown in figure 4.

Process Improvement

Improving process quality requires sustained efforts. It is a commitment to adopting preventive solutions using a systematic approach. The variability reduction technique is one systematic approach to problem solving. The process generally follows this sequence:

1. Describe the process and break it into subprocesses.
2. Determine measurement instrument variability.
3. Collect and analyze data.
4. Identify major component of variation using multivary chart.
5. List input variables associated with the major component of the variation. Screen out trivial variables using brainstorming, screening experiments, or similar techniques.
6. Conduct statistically designed experiments to identify critical input variables.
7. Optimize the process with attention toward critical variables and determine realistic tolerances of input variables.
8. Control critical input variables within the realistic tolerance.
9. Reevaluate process variability. If necessary, repeat steps 1 through 8.

Running a statistically designed experiment requires more than just knowledge of the statistical techniques. Familiarity with the manufacturing environment, knowledge of the process, and collection of appropriate data contribute significantly toward the successful completion of the experiment. A step-by-step process to run an experiment is described below.

inherent capability of a process with respect to the specification limits, while the process performance, C_{pk}, is a measure of how well a process is doing over time with respect to the process target (i.e., shift in process mean). Mathematically, the C_p and C_{pk} are defined as follows:

C_p = Specification Width (USL-LSL) / Process Width (6 Sigma)

and,

$C_{pk} = (1-k)C_p$ where K = (difference between Process Mean and Target) / Half the Tolerance (USL-LSL)/2; and

C_{pk} = Minimum of [(USL-$\bar{X}$) or ($\bar{X}$-LSL)]/3 * Sigma.

where USL and LSL represent the upper specification limit and the lower specification limit, respectively. For a centered process, $C_p = C_{pk}$, otherwise $C_{pk} < C_p$. Usually, the acceptable value of C_p is 1.33. However, superior companies, like Motorola, IBM, or DEC have acceptable values of C_p = 2. These two parameters (C_p and C_{pk}) are becoming the industry standard for measuring relative process quality. With the use of process variability as a measure of process quality, the focus shifts from a "within specification" to an "on target" philosophy. The goal becomes a reduction in variability around the process target rather than an attempt to contain the output within specification by sorting.

There are several tools available to analyze data and to display variation. The most effective tools are the cause-and-effect diagram and multivary charts. Some of the other tools are Pareto

1. State the objective of the experiment.
2. Brainstorm the experiment within a cross-functional team.
3. Identify output characteristics and the expected responses.
4. List input variables and determine operating range for the experiment.
5. Determine experimental technique to be used (i.e., comparative experiments, factorial designs, Taguchi methods or others).
6. Design the experiment (i.e., layout of the experiment, sample size, number of replications, and analysis method).
7. Document procedure for running the experiment. Include instructions for any special information, check sheets, or emergency situations.
8. Run the experiment through normal manufacturing conditions rather than special conditions. Experiments do not require "baby-sitting."
9. Monitor progress throughout the experiment and record observations.
10. Analyze the data and determine if a confirmation experiment is necessary.
11. Make recommendations for process changes.
12. Document the experiment design, analysis, results, and recommendations for future use.

Some of the widely used experimental designs include the one-variable-at-a-time approach, comparative experiments (using t-test, F-test, Chi-Square test, and other nonparametric methods), factorial designs, orthogonal arrays, and other techniques. The strategy for running experiments (figure 5) shows that full factorial experiments are suitable when the number of variables is less than six. Though factorial experiments are usually run at two levels, they can be run at multiple levels. Running a four factor, two-level experiment requires a total of 16 cells or treatments. The total number of experimental cells, C, are given by $C = L^V$, where L is the number of levels and V is the number of variables. A two-level, two-variable full factorial design is shown in figure 6 for reducing solder defects. With the use of a full factorial design one can determine the main effects of the belt speed, the solder temperature, and interaction between them.

The experiment can be analyzed using a table of contrast or Yates algorithm.[3] For a larger experiment, the significance of factors can be analyzed using normal probability paper or analysis of variance (ANOVA).[4] For this experiment, the table of contrast and summary of the results are shown in figure 7. Results can also be displayed graphically for visualization of the responses.

	Belt Speed	
Temperature	Low	High
Low	A1B1 15	A2B1 25
High	A1B2 11	A2B2 23

Figure 6, a 2 x 2 full factorial design for reducing solder defects.

	Mean	A	B	AB	Results
	+	−	−	+	15
	+	−	+	−	11
	+	+	−	−	25
	+	+	+	+	23
All (+)s	74	48	34	38	
All (−)s	0	26	38	36	
Effects (delta)	74	22	4	2	
Avg. Effect	18.5	11	2	1	

Figure 7, table of contrasts for 2 x 2 experiments. Results can be displayed graphically for visualization of responses.

Process Optimization

Optimization establishes relationships between output and critical input variables. The objective of optimization is to find parameter settings that produce maximum yield. Response surface methodology (RSM) is a sequential approach that uses factorial designs to visualize the problem and responses of the variables and lead to process optimization.

Advanced optimization techniques for many variables (in the hundreds or thousands) utilize linear programming algorithms run on mainframe computer systems. Such algorithms are mainly used for effective resource utilization.

Process Maintenance

Process maintenance can be handled by using in-line inspection or on-line control charts, which help to maintain process output within specification limits. A control chart is a statistical tool used to differentiate between random and assignable variations. The random variation is inherent in the process while the assignable variation is attributed to specific causes. Attribute control charts use discrete measurements such as good or bad, or percent defective. Widely used attribute control charts are a P-chart for fraction defective, a U-chart for number of defects per unit, and a C-chart for number of defects. Variable control charts use measurement data on a continuous scale such as voltage or temperature. Most widely used variable control charts are X-bar and R charts. The process steps to establish a control chart: 1. identify the characteristics to be charted, 2. determine the type of control charts, 3. decide the target line to be used, 4. choose the sample size and frequency, 5. set up the system for collecting the data, 6. calculate the control limits,[5] and 7. document instructions on interpretation of the results and actions to be taken.

For a process to be out-of-control, at least one of the following conditions must occur.
1. One of more points outside the control limits.
2. A run of seven or more points, establishing a trend.
3. Nonrandom patterns in the data.
4. Two or three points outside of two sigma limits.
5. Four or five points on or beyond one sigma limits.

It must be remembered that successful implementation of control charts can only be achieved by knowing the theory behind them. **smt**

References

1. National Institute of Standards and Technology, "1990 Application Guidelines — Malcolm Baldrige National Quality Award," 1990.
2. Ishikawa, K., "Guide to Quality Control," Asian Productivity Organization, 1982.
3. Box, George E., Hunter, William G., and Hunter, J. Stuart, "Statistics for Experimenters," John Wiley & Sons, Inc., 1978.
4. Duncan, Acheson J., "Quality Control and Industrial Statistics," Richard D. Irwin, Inc., 1986.
5. Juran, J.M. and Gryna, Frank M., "Juran's Quality Control Handbook," McGraw-Hill Book Co., 1988.
6. Delott, Charles and Gupta, Praveen, "Characterization of Copper Plating Process for Ceramic Substrates," *Quality Engineering,* 2(3), 269-284, 1990.
7. Deming, W.E., "Out of the Crisis," MIT Center for Advanced Engineering Study, 1982.

Contact the author at Quality Technology Co., 1270 Lance Lane, Carol Stream, Illinois 60188; telephone: 708/ 231-3142.

Experiments with Three Factors: 2^3

It is time to move on to the next size DOE: one using three factors. The same analytical procedures are used. With three factors in a 2^3 factorial, there are eight runs or combinations of the three factors. From these basic eight runs can be computed the analysis columns for three main effects, three 2-factor interactions, and one 3-factor interaction. Although the 3-factor interaction is discussed in this chapter, it is usually ignored since such interactions rarely occur. (The exceptions to this statement, in this author's experience, have been with some chemical processes.)

Example 3: Chemical-Processing Yield

A team is striving to improve the yield of a chemical process. The controlling factors are hypothesized to be the processing temperature, the concentration level of a catalyst, and the ramp time (the time it takes to go from room temperature to the processing temperature). A formulation is processed for a specified time, cooled, and dissolved in a reagent. Then the end-product is precipitated. The levels chosen for this study are

A: Temperature—400°F and 450°F

B: Concentration—10 percent and 20 percent

C: Ramp time—45 seconds and 90 seconds

The response is the yield of the refined product as a percentage of the raw stock. The objective is to maximize the yield. The design is a 2^3 factorial. Each run is replicated three times with the twenty-four trials randomized as to order of test. The data are shown in Table 3.1.

TABLE 3.1. Data for Example 3.

		A: Temperature			
		(−)		(+)	
		C: Ramp Time		C: Ramp Time	
		(−)	(+)	(−)	(+)
B: Conc.	(−)	66.63 62.01 (− − −) 57.85	60.31 60.87 (− − +) 63.93	77.25 70.33 (+ − −) 67.73	69.98 67.28 (+ − +) 67.54
	(+)	50.25 59.95 (− + −) 56.05	56.46 58.03 (− + +) 54.72	66.91 70.16 (+ + −) 74.67	74.88 73.12 (+ + +) 73.80

The first step in analyzing this experiment is to convert the data table into a spreadsheet format. Table 3.2 should be verified as an exercise. Extreme care is required to assure that the run averages and variances are placed in the correct rows. There is no recovery from an error! (The experimenter, if confident of his or her data identification, could bypass the data table and go directly to the spreadsheet.)

TABLE 3.2. Example 3 in spreadsheet format.

Run	A	B	C	AB	AC	BC	ABC	$\bar{Y}$	S^2
1	−	−	−	+	+	+	−	62.16	19.27
2	+	−	−	−	−	+	+	71.77	24.21
3	−	+	−	−	+	−	+	55.42	23.81
4	+	+	−	+	−	−	−	70.58	15.21
5	−	−	+	+	−	−	+	61.70	3.80
6	+	−	+	−	+	−	−	68.27	2.22
7	−	+	+	−	−	+	−	56.40	2.76
8	+	+	+	+	+	+	+	73.93	.79
ΣY_+	284.55	256.33	260.33	268.37	259.78	264.26	262.82		
ΣY_-	235.68	263.90	259.93	251.86	260.45	255.97	257.41		
$\bar{Y}_+$	71.14	64.08	65.08	67.09	64.95	66.07	65.71		
$\bar{Y}_-$	58.92	65.98	64.98	62.97	65.11	63.99	64.35		
Effect	12.22	−1.90	0.10	4.12	−0.16	2.08	1.36		

Use the eight-step analysis procedure.

Step 1. Calculate main effects and interaction effects from the analysis table (Table 3.2).

Step 2. Create the Pareto chart of effects (see Figure 3.1).

Step 3. Calculate S_e.

$$S_e = \sqrt{(\Sigma S_i^2 / k)}$$
$$= \sqrt{[(19.27 + 24.21 + 23.81 + 15.21 + 3.80 + 2.22 + 2.76 + .79)/8]}$$
$$= \sqrt{11.51} = 3.39$$

Step 4. Calculate S_{eff}.

$$S_{eff} = S_e \sqrt{(4/N)} = 3.39 \, (\sqrt{(4/24)}) = (3.39)(.41) = 1.38$$

Step 5. Determine degrees of freedom and t-statistic.

$$df = (\text{\# replicates per run} - 1)(\text{\# of runs}) = (2)(8) = 16$$
$$t_{df=16; \alpha=.05} = 2.12$$

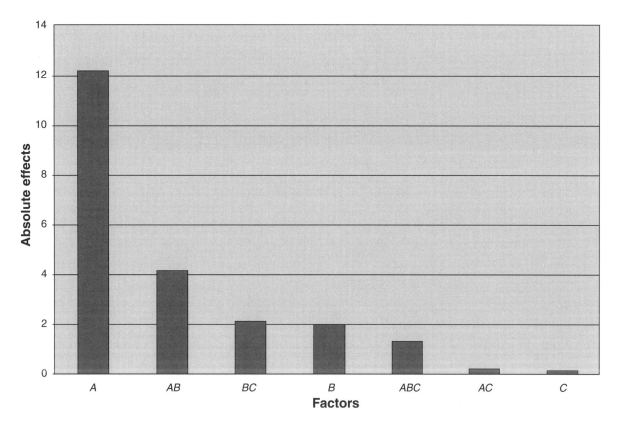

FIGURE 3.1. Pareto of effects.

Step 6. Compute decision limits.

$$DL = \pm tS_{eff} = \pm (2.12)(1.38) = \pm 2.93$$

As the comparison of the effects against the decision limits shows, only A (temperature) and AB (temperature-concentration interaction) are significant.

Step 7. Plot the significant effects. The significant main effect, temperature, also has a significant interaction with concentration. One need only plot the interaction since the significant interaction means that both factors must be considered to evaluate the response. Interaction plots require generating a 2×2 table of means using the four pairs of signs of the two factors from the analysis table $(--, -+, +-, ++)$. The cell means are grouped as in Table 3.3.

TABLE 3.3. Interaction table for Example 3.

		Temperature (−)	Temperature (+)
Concentration	(−)	62.16 61.70 $\overline{Y} = 61.93$	71.77 68.27 $\overline{Y} = 70.02$
	(+)	55.42 56.40 $\overline{Y} = 55.91$	70.58 73.93 $\overline{Y} = 72.26$

This table can be used to plot the interaction as either *AB* or *BA*, as illustrated in Figures 3.2 and 3.3. They are equivalent, but sometimes one plot communicates better than the other. The experimenter should try both to see which *speaks* to him or her more clearly.

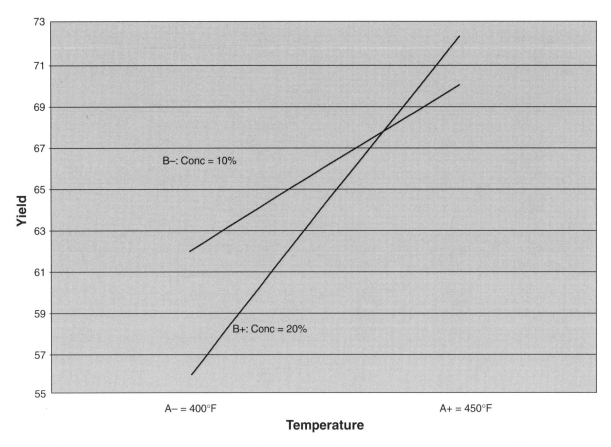

FIGURE 3.2. Temperature-concentration interaction.

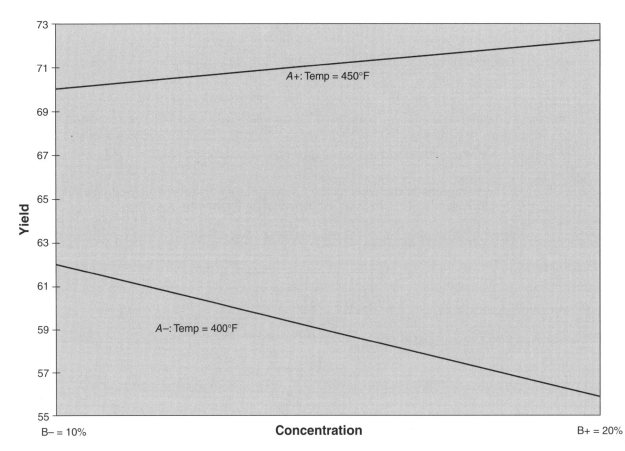

FIGURE 3.3. Concentration-temperature interaction.

Step 8. Determine model and conclusions.

These results indicate that yield is maximized at the high level of temperature and the high level of concentration. The ramp time did not matter nor did any other interactions. This implies ramp time can be set solely on the basis of cost, productivity, or convenience. The final question is what yield can be expected at the proposed optimum condition. This is obtained from the model. Remember that the hierarchy rule requires that the main effects of significant interactions must be included in the model even when they are not significant. Therefore, the main effect for factor B is included even though it was not significant.

$$\hat{Y} = \overline{\overline{Y}} + (E(A)/2)A + (E(B)/2)B + (E(AB)/2)AB =$$
$$\hat{Y} = 65.03 + 6.11A - .95B + 2.06AB$$
$$= 65.03 + (6.11)(+1) - (.95)(+1) + (2.07)(+1)(+1) = 72.26.$$

It would be prudent to confirm these results by operating at the proposed condition and verifying that the forecast results are obtained. One might also consider further experimentation. If increasing the temperature by 50° was beneficial with 20 percent concentrate, can the temperature be increased further to increase yield? Will an even higher concentrate level improve the yield even further? The next experiment is already fairly clear. The point to remember is that most problems are solved by a sequence of studies that move toward an optimum region. This requires that one be aware of physical and safety constraints that may limit further studies. If still not satisfied with the performance, look for other factors.

Variation Analysis

Return to Table 3.1. Has all the relevant and useful information been squeezed from this experiment? Does anything else attract one's attention as one reviews the raw data? After a few minutes of studying the raw data, go to the analysis table (Table 3.2) and review the summaries listed there. What are the objectives of a DOE? They are to

1. Control or move the process average
2. Reduce variation
3. Make the process more robust

The study of effects addresses both objectives 1 and 3. Was any pattern noted in the variances of the runs? A procedure is needed to address objective 2, the reduction of variation.

Now determine if any of these factors (or their interactions) will influence variation. This is done by analyzing the variances of the cells as the *response* for a variation analysis. The *analysis procedure for variation* follows.

1. Calculate the variances (S_i^2) of each run or treatment.
2. Calculate the average variance for the high level ($\overline{S}_+^2$) and low level ($\overline{S}_-^2$) of each factor and interaction.
3. For each factor and interaction, calculate a new statistic:
   $F = (\overline{S}^2_{larger} / \overline{S}^2_{smaller})$.
4. For a given risk level (generally 10 percent), go to an *F-table* for the critical test value. (Refer to Table A.2 in the Appendix.) An F-table is indexed by the risk level and the degrees of freedom of the numerator and denominator terms. In all of the problems contained in this text, the degrees of freedom for the numerators and denominators are equal. That is, the degrees of

freedom calculated for the t-statistic in step 5 in the analysis of effects are divided in half for the numerator variance and for the denominator variance. This provides the test statistic for evaluating significant differences in variances. If the calculated F is greater than the F-table value, the differences are significant and a factor that influences variation appears to have been found. (There are some complicated issues being ignored here. Since this F-test is based on placing the larger variance over the smaller variance, this is a *two-tailed test* with a 5 percent alpha risk in each tail. Thus, the risk is actually 10 percent, and Table A.2 is based on 5 percent risk in each tail. For further study, the concepts of one-tailed and two-tailed risk should be reviewed in a basic statistics text.)

5. Evaluate the conclusions for variation. Determine optimum operating conditions for both effects and variation. If the two objectives conflict, a compromise must be reached.

Table 3.2 can be expanded (as shown in Table 3.4) using the additional tools for variation analysis.

TABLE 3.4. Expansion of Example 3 for variation analysis.

Run	A	B	C	AB	AC	BC	ABC	$\bar{Y}$	S^2
1	−	−	−	+	+	+	−	62.16	19.27
2	+	−	−	−	−	+	+	71.77	24.21
3	−	+	−	−	+	−	+	55.42	23.81
4	+	+	−	+	−	−	−	70.58	15.21
5	−	−	+	+	−	−	+	61.70	3.80
6	+	−	+	−	+	−	−	68.27	2.22
7	−	+	+	−	−	+	−	56.40	2.76
8	+	+	+	+	+	+	+	73.93	.79
ΣY_+	284.55	256.33	260.30	268.37	259.78	264.26	262.82		
ΣY_-	235.68	263.90	259.93	251.86	260.45	255.97	257.41		
$\bar{Y}_+$	71.14	64.08	65.08	67.09	64.95	66.07	65.71		
$\bar{Y}_-$	58.92	65.98	64.98	62.97	65.11	63.99	64.35		
Effect	12.22	−1.90	0.10	4.12	−0.16	2.08	1.36		
S^2_+	10.61	10.64	2.39	9.77	11.52	11.76	13.15		
S^2_-	12.41	12.38	20.63	13.25	11.50	11.26	9.87		
F	1.17	1.16	8.63	1.36	1.00	1.04	1.33		

The original degrees of freedom were 16. For a 5 percent risk in the upper tail (10 percent total), refer to the F-table (Table A.2, Appendix) for a test statistic based on 8 degrees of freedom for numerator and denominator respectively: $F_{(df1=8;\, df2=8)} = 3.44$. Only one factor (C: ramp time) has an F-value exceeding that test statistic. Since the F is based on the larger variance divided by the smaller, one must check the individual variances to see that the high (+) ramp time level will provide lower variation.

The experimenter's conclusions must now be modified. Based on effects alone, there was no reason to be concerned about the level of ramp time because it was not significant. Adding the variation analysis, it is now apparent that it is quite important to have a longer ramp time in order to reduce the variation in the yield. These conclusions have a major impact on the estimate of inherent variation for the predictions from the model. If ramp time is at the higher (+) level, the inherent variation under those conditions is estimated by $S_e = \sqrt{2.39} = 1.55$. If the lower level (−) of ramp time is used, the responses have an estimate of inherent variation of $S_e = \sqrt{20.63} = 4.54$. Estimation of spread or C_{pk}s are impacted by which standard deviation is used. More importantly, the inherent variation in the process is affected.

Some general guidelines for this joint analysis of effects and variation follow:

1. Significance of effects is independent of the assessment of variation, and vice versa. They are two uncorrelated assessments.

2. It is possible to be unlucky and have a *conflict*; i.e., a factor may occasionally be significant for both effects and variation, with different levels required for the optimization of each. This requires a compromise or trade-off in trying to achieve an adequate, if not ideal, operating condition for both objectives.

3. Variation analysis is particularly susceptible to *outliers* or special cause variation that results from influences outside the DOE. These occurrences distort the assessment of variation. An indicator of this problem is the occurrence of more than one significant term in the variation analysis.

4. If an effect that will control process averages is worth its weight in gold (a *golden* factor), then a factor that controls variation is a *diamond* factor. Such factors are relatively rare, and, when they occur, they may be the most important discovery in an experiment. Generally, factors can be found to change the averages whereas it is difficult to find a factor that reduces variation.

This is the completion of the analysis procedure; it should be used in its entirety in subsequent exercises. When an experiment is replicated (two or more trials for each cell), the assessment of variation is a free extension of the analysis since the data are already available without additional testing required.

Exercise 3: Ink Transfer

Analyze the following experiment for effects and for variation. State conclusions and recommendations for best operating conditions. Refer to the analysis in the appendix when you have completed your work.

In an experiment to evaluate ink transfer when printing on industrial wrapping paper, the following three factors were used in a factorial design:

Factor	Low Level (−)	High Level (+)
A: Roll Type	Light	Heavy
B: Paper Route	Over Idler Roll	Direct
C: Drying	Fan Only	Fan Plus Heater

34 AN INTRODUCTION TO DESIGN OF EXPERIMENTS

The ink transfer was measured by averaging inspector ratings during the processing of a roll of paper. The objective was to minimize this rating. Three independent rollings were made for each run. The order of the 24 trials was randomized. The analysis table is provided in Table 3.5.

TABLE 3.5. Analysis table for ink transfer exercise.

Run	A	B	C	AB	AC	BC	ABC	$\bar{Y}$	S^2
1	−	−	−	+	+	+	−	10.00	1.00
2	+	−	−	−	−	+	+	18.33	4.35
3	−	+	−	−	+	−	+	11.5	1.50
4	+	+	−	+	−	−	−	20.00	1.80
5	−	−	+	+	−	−	+	5.00	12.50
6	+	−	+	−	+	−	−	21.67	11.75
7	−	+	+	−	−	+	−	13.33	9.50
8	+	+	+	+	+	+	+	23.33	7.87

ΣY_+
ΣY_-
$\bar{Y}_+$
$\bar{Y}_-$
Effect
$\bar{S}_+^2$
$\bar{S}_-^2$
F

Analysis with Unreplicated Experiments

What can be done for analysis if there is no replication? This is a common occurrence in DOE due to the expense of the experimentation. The usual procedures will not work since there are no variances without replication. Consider the abbreviated analysis table in Table 3.6 for an unreplicated experiment with three factors (2^3).

TABLE 3.6. Variation analysis in an unreplicated experiment.

Run	A	B	C	AB	AC	BC	ABC	Y
1	−	−	−	+	+	+	−	15.1
2	+	−	−	−	−	+	+	22.1
3	−	+	−	−	+	−	+	5.0
4	+	+	−	+	−	−	−	16.9
5	−	−	+	+	−	−	+	14.3
6	+	−	+	−	+	−	−	23.2
7	−	+	+	−	−	+	−	9.1
8	+	+	+	+	+	+	+	21.9
Effect	10.15	−5.45	2.35	2.20	.70	2.20	−.25	

The experimenter cannot evaluate the significance of the effects since there is no estimate of S_e. Consider the Pareto of the effects in Figure 3.4.

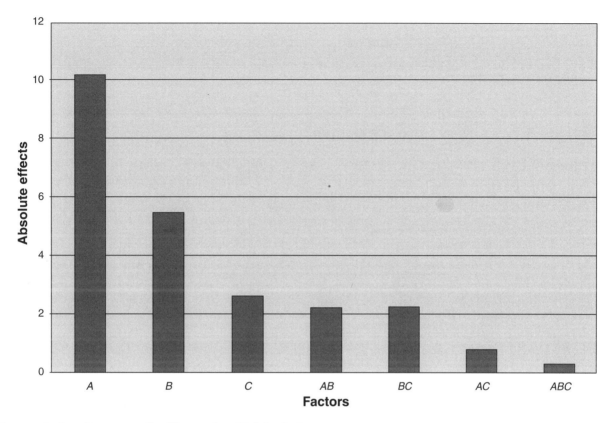

FIGURE 3.4. Pareto of effects for Table 3.6.

In the absence of the more formal procedure, the *eyeball* test here indicates that factor A and factor B are quite probably realistic effects. Assume that they can be used in a model. (A more formal name for the Pareto analysis is a *scree* analysis. When switching to software, there is an equivalent technique using probability plots of effects.) The model is

$$\hat{Y} = 15.95 + [E(A)/2]A + [E(B)/2]B$$

and

$$\hat{Y} = 15.95 + 5.07A - 2.72B$$

The appropriate plots could be made and conclusions drawn.

What about variation analysis? Following is a variation of a technique discussed in Montgomery* (1991) that can be applied even though there are no variances to compare.

1. Calculate the *predicted* (*pred*) response for each run by using the model and enter in the column for predictions. For the first row, $A = -1$, and $B = -1$. For that row, the model predicts

$$\hat{Y} = 15.95 + 5.07A - 2.72B = 15.95 - 5.07 + 2.72 = 13.60$$

*From Montgomery, Douglas. 1991. *Design and Analysis of Experiments*. New York, NY: J. Wiley & Sons.

2. Calculate a *residual* (*res*) for each run by calculating

$$\text{Residual} = Y - \text{predicted}$$

and enter the results in the last column of the table. For the first row,

$$\text{Residual} = 15.1 - 13.6 = 1.5.$$

These results are shown in Table 3.7.

TABLE 3.7. Analysis table for residuals.

Run	A	B	C	AB	AC	BC	ABC	Y	Pred	Res
1	−	−	−	+	+	+	−	15.1	13.6	1.5
2	+	−	−	−	−	+	+	22.1	23.7	−1.6
3	−	+	−	−	+	−	+	5.0	8.2	−3.2
4	+	+	−	+	−	−	−	16.9	18.3	−1.4
5	−	−	+	+	−	−	+	14.3	13.6	.7
6	+	−	+	−	+	−	−	23.2	23.7	−.5
7	−	+	+	−	−	+	−	9.1	8.2	.9
8	+	+	+	+	+	+	+	21.9	18.3	3.6
S_+^2	5.91	8.66	3.00	4.27	8.40	4.58	8.73			
S_-^2	4.60	1.85	3.83	3.01	1.78	2.70	1.74			
F	1.28	4.68	1.28	1.42	4.72	1.70	5.02			

3. For each column, calculate the variance of the residuals for the high (+) level of the column and then for the low (−) level of the column. Enter in the rows at the bottom of the table. As an example, the residuals associated with pluses in the A column are −1.6, −1.4, −.5, 3.6 and $S_+^2 = 5.91$.

4. Calculate an F-statistic for each column by dividing the larger variance by the smaller variance. For each variance, the degrees of freedom are the number of residuals used to calculate a variance minus one; i.e., the degrees of freedom=3 for each variance in this case.

5. Using the F-table (Table A.2, Appendix) for 10 percent risk and $df = (3,3)$, determine the test value for comparison. Any factor or interaction whose F statistic exceeds the test value has significant influence on variation. In this instance, F-table value = 9.28. Since none of the F-tests exceed that test value, none of the factors are significant with respect to their influence on variation.

While this approach looks tedious, remember that it is actually *free* information; that is, it requires no additional data. When referring to software, this technique can be easily applied with a spreadsheet and it is automatically calculated with specialized software. This technique is weak (will only detect very large differences in variances) when the total degrees of freedom are less than 14. With larger experiments, it is a very useful tool.

Screening Designs

The necessary analytical techniques have now been covered. They will work for any number of factors in a DOE as long as only two levels are used. As the number of factors increases in a factorial design, the cost and difficulty of control increase even faster due to the acceleration in the number of trials required. It's necessary now to study the design techniques that economically permit DOEs with many factors. These are the *screening* designs that can be used for as few as four factors and as many as can be practically handled or afforded in an experiment. Their purpose is to identify or screen out those factors that merit further investigation. To provide some idea of the power that is being introduced, consider Table 4.1.

TABLE 4.1 Comparison of the number of runs in factorial and screening designs.

	# of Runs in a DOE	
# of Factors	Full Factorial	Minimum Screening Design
4	16	8
5	32	8
6	64	8
7	128	8
11	2,048	12
15	32,768	16
19	524,288	20
23	8,388,68	24
27	134,217,728	28
31	2,147,483,648	32

Can you imagine the budget for a 2^{31} factorial experiment? Now *that* is job security! If you performed one experiment per minute, 24 hours a day, 365 days per year, it would take 4,085.8 years to complete the DOE! Even a much smaller 2^7 factorial requires 128 trials—still an expensive investment for most processes. The comparative numbers required by screening designs are not typographical errors; they are the smallest experiment that can provide substantial information. Looks too good to be true? You are right—there are penalties to be paid for this economy. But the penalties are minor compared to the immense advantages to be gained.

Before going further, the concepts of *confounding* and *aliases* must be defined. Consider the analysis matrix for a factorial experiment with three factors (Table 4.2).

TABLE 4.2. Analysis table for three factors.

A	B	C	AB	AC	BC	ABC ⇓ D
−	−	−	+	+	+	−
+	−	−	−	−	+	+
−	+	−	−	+	−	+
+	+	−	+	−	−	−
−	−	+	+	−	−	+
+	−	+	−	+	−	−
−	+	+	−	−	+	−
+	+	+	+	+	+	+

Assume that there actually are *four* factors to be studied requiring 16 runs for a full factorial. Assume further that, due to material limitations or cost, only eight runs can be performed. How could an extra factor be packed into this design? Why not put factor D in the ABC column and treat that column like a factor? It has already been said that a three-factor interaction isn't realistic, so it probably wouldn't be missed. What are the penalties of this move? The penalty is that each two-factor interaction is now identical with another two-factor interaction. For example, calculate a column for CD and compare it to the AB column. They are identical, and $AB = CD$. This is illustrated in Table 4.3.

TABLE 4.3. Illustration of identical interactions.

A	B	AB	C	D	CD
−	−	+	−	−	+
+	−	−	−	+	−
−	+	−	−	+	−
+	+	+	−	−	+
−	−	+	+	+	+
+	−	−	+	−	−
−	+	−	+	−	−
+	+	+	+	+	+

When the AB effect is calculated, the experimenter doesn't know if that value is the AB effect, the CD effect, or a combination of both. This confusing of effects is called *confounding*, and it is said that AB is confounded with CD. The confounded effects are also called *aliases*; aliases are identical except for sign. If AB and CD are aliases, their sign column in an analysis matrix can be either identical or opposite in sign. The same is true of their calculated effects. The decision to treat factor D as if it was the ABC column generates an *identity* that can be used to define the aliases. That is, the aliases are defined as $D = ABC$. The rule for manipulating such relationships follows.

For any defined relationship, *a squared term will equal unity or 1*; e.g., if both sides of the previous relationship are multiplied by *D*, the result is

$$D \times D = D^2 = 1 = ABCD.$$

This is the defining relationship or *identity* that defines the aliases when a fraction of the full factorial design is performed. Generally, the *1* is replaced with an *I*. To determine the aliases for any effect, multiply both sides of the identity by that effect. For this example, the identity is

$$I = ABCD$$

Multiplying both sides by *AB* will provide the alias for *AB*:

$$IAB = AB = A^2 B^2 CD = CD$$

Thus, *AB* is confounded with *CD*; one can also state that *AB* and *CD* are aliases.

Table 4.4 shows all the aliases for this design, which is a $\frac{1}{2}$ fractional factorial ($\frac{1}{2}$ the number of runs for a full factorial).

TABLE 4.4. Aliases for a $\frac{1}{2}$ fractional factorial design with four factors.

A = BCD	AB = CD
B = ACD	AC = BD
C = ABD	AD = BC

As more factors are packed into a design, the definition of the identities becomes more difficult. Therefore, the aliases are provided or described in all of the screening designs used by this author. Confounding is the penalty paid for using fewer runs than a full factorial in a design. While there are times that this penalty is very troublesome, most of the time it is a risk that can be managed and rules are covered to reduce this risk further. In any screening design, one must be aware of the confounding pattern.

Next, the *intent* of a screening design must be reviewed. Such designs are for sifting through a large number of factors and selecting those that merit more detailed study. In screening designs, the primary interest is main effects. The levels are set broadly to increase the chance of detecting significant factors. If interactions exist, it is great if they can be identified, but the most important consideration is that they not mislead identification of the factors for further study. The reason for this concern is that in the larger screening designs (five factors and larger), main effects will be confounded with two-factor interactions, either completely or partially. A technique for separating main effects from this confounding with two-factor interactions is discussed later, but even then two-factor interactions remain confounded with one another.

Recall that the order of studying designs was said to be backward. For ease of instruction, the author started with small numbers of factors and progressed to large numbers of factors. In practice, a study begins with a screening design. It is the tool used to sift through many potential factors, identifying those that need to be studied in smaller refining experiments in which interactions can be more thoroughly assessed.

Screening designs can be used to study n-1 *factors in* n *runs in which* n *is divisible by four.* That is, 7 factors can be studied in 8 runs, 11 factors can be studied in 12 runs, 15 factors in 16 runs, etc. Such designs are of two distinctly different types:

1. *Geometric designs.* These designs are those in which *n* is a power of two. That is, the number of runs is 8, 16, 32, 64, etc. These designs may be *fractional factorials* or *Plackett-Burman* designs.

They are the same if a maximum number of factors is used in the design. In the geometric designs, the confounding that exists is complete. In other words, if two effects are confounded, they are identical except possibly for sign. (If *AB* and *CD* are confounded, their effects are numerically identical but one may be the negative of the other.) It is easy to specify precisely what is confounded with what (the aliases), and it should be done. This perfect confounding is the advantage of geometric designs in that one can identify the risks in the interpretation of effects.

2. *Nongeometric designs.* These are designs that are divisible by four but are not powers of two. Such designs have runs of 12, 20, 24, 28, 36, etc. These designs are Plackett-Burman designs and represent a unique contribution to the available designs. These designs *do not have* complete confounding of effects. Instead, each main effect is confounded partially with all interactions that do not contain that main effect. This is both an advantage and a disadvantage. The advantage is that all of this partial confounding tends to average out to a negligible impact, and it may be possible to get some information on a significant interaction. The disadvantage is that presence of a large interaction may distort the estimated effects of several individual factors since each interaction is partially confounded with all main effects except the two interacting factors.

This discussion touches the tip of a technical iceberg that is deferred to the study of more advanced references. One should note that while fractional factorial designs are more difficult to portray, software makes them very easy to use. Fractional factorials assure efficient control of the confounding pattern whereas Plackett-Burman designs are easiest to apply. This comparison becomes a concern only when a design is performed with less than the maximum number of possible factors. Plackett-Burman designs are provided for six or more factors. Fractional factorial designs are provided for four- and five-factor experiments in which a full factorial is not practical. (Special designs are provided for these two cases since there is better efficiency in using fractional factorials for them than in using Plackett-Burman.) Designs are listed in the Appendix. There are *two critical characteristics of screening designs* that help make the screening designs work despite confounding:

1. *Effect heredity.* Large main effects often have significant interactions. This provides guidelines on where to look for interactions.

2. *Effect sparsity.* Main effects are generally much larger than two-factor interactions, while three-factor and higher order interactions can generally be ignored. This means that main effects generally dominate interactions, and it allows the experimenter to ignore three-factor (and higher) interactions as rare events.

The eight-step analysis procedure for effects and the variation procedure are applied in screening designs just as they have been applied in previous examples.

Example 4: An Eight-Run Plackett-Burman Design with Seven Factors

A manufacturer of paperboard products needed to increase the puncture resistance of his product. The response was the force required to penetrate the material. Brainstorming with a team from the shop resulted in seven factors for inclusion in the DOE, with levels in order of low vs. high:

A: Paste temperature—130°F vs. 160°F

B: Additive in paper stock to inhibit moisture—.2 percent vs. .5 percent

C: Press roll pressure—40 psi vs. 80 psi

D: Paper moisture—low vs. high

E: Paste type—no clay vs. with clay

F: Cure time for finished paperboard—10 days vs. 5 days

G: Machine speed—120 fpm vs. 200 fpm

An eight-run Plackett-Burman design is selected for the DOE, and two replicates are generated for each run. The objective is to maximize the response. The design matrix, the data, and the aliases are shown in Table 4.5.

TABLE 4.5. Example 4.

A	B	C	D	E	F	G	Y1	Y2
+	−	−	+	−	+	+	12.5	16.84
+	+	−	−	+	−	+	42.44	39.29
+	+	+	−	−	+	−	55.08	47.57
−	+	+	+	−	−	+	49.37	47.69
+	−	+	+	+	−	−	55.43	52.80
−	+	−	+	+	+	−	42.51	35.02
−	−	+	−	+	+	+	51.13	57.92
−	−	−	−	−	−	−	15.61	13.65
−BD	−AD	−AG	−AB	−AF	−AE	−AC		
−CG	−CE	−BE	−CF	−BC	−BG	−BF		
−EF	−FG	−DF	−EG	−DG	−CD	−DE		

For example, the aliases of factor A mean that

$$E(A) = -E(BD) = -E(CG) = -E(EF)$$

The analysis proceeds as usual; it is not affected by the size of the experiment. Verify the contents of the analysis table, Table 4.6.

TABLE 4.6. Analysis table for Example 4.

Run	A	B	C	D	E	F	G	$\bar{Y}$	S^2
1	+	−	−	+	−	+	+	14.67	9.42
2	+	+	−	−	+	−	+	40.87	4.96
3	+	+	+	−	−	+	−	51.33	28.20
4	−	+	+	+	−	−	+	48.53	1.41
5	+	−	+	+	+	−	−	54.12	3.46
6	−	+	−	+	+	+	−	38.77	28.05
7	−	−	+	−	+	+	+	54.53	23.05
8	−	−	−	−	−	−	−	14.63	1.92
ΣY_+	160.99	179.50	208.51	156.09	188.29	159.30	158.60		
ΣY_-	156.46	137.95	108.94	161.36	129.16	158.15	158.85		
$\bar{Y}_+$	40.25	44.88	52.13	39.02	47.07	39.83	39.65		
$\bar{Y}_-$	39.12	34.49	27.24	40.34	32.29	39.54	39.71		
Effect	1.13	10.39	24.89	−1.32	14.78	.29	−.06		
$\bar{S}_+^2$	11.51	15.66	14.03	10.59	14.88	22.18	9.71		
$\bar{S}_-^2$	13.61	9.46	11.09	14.53	10.24	2.94	15.41		
F	1.18	1.65	1.27	1.37	1.45	7.54	1.59		

1. Effects are shown in the analysis table. Verify the calculations of at least two effects.
2. The Pareto chart of effects is represented in Figure 4.1. Decisions on significance could be made solely from the Pareto chart with only modest increase in the risk of an erroneous decision. Based on the Pareto chart, it is obvious that C, E, and B are dominant.

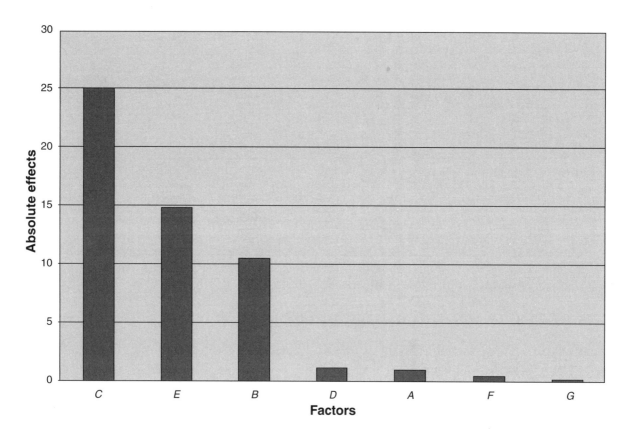

FIGURE 4.1 Pareto of effects for Example 4.

3. Standard deviation of the experiment.

$$S_e = \sqrt{(\Sigma S_i^2)/k} = \sqrt{(9.42 + 4.96 + \ldots + 1.92)/8} = \sqrt{12.56} = 3.54$$

4. Standard deviation of the effects.

$$S_{eff} = S_e \sqrt{(4/N)} = 3.54\sqrt{(4/16)} = 1.77$$

5. Degrees of freedom and t-statistic.

$$df = (\text{\# replicates} -1) \times (\text{\# of runs}) = (2-1)(8) = 8$$

and

$$\alpha = .05 \Rightarrow t_{df=8} = 2.31$$

6. Decision limits (see Figure 4.2).

$$DL = \pm t \, S_{\text{eff}} = \pm (2.31)(1.77) = \pm 4.09$$

Based on the decision limits, factors C (press roll pressure), E (paste type), and B (additive level) are significant and influence the response.

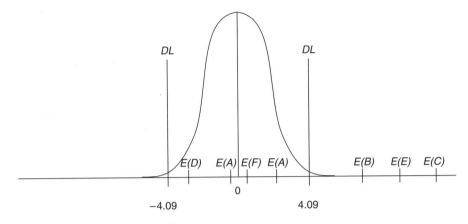

FIGURE 4.2 Decision limits and effects.

7. Plots of significant effects are shown in Figures 4.3, 4.4, and 4.5.

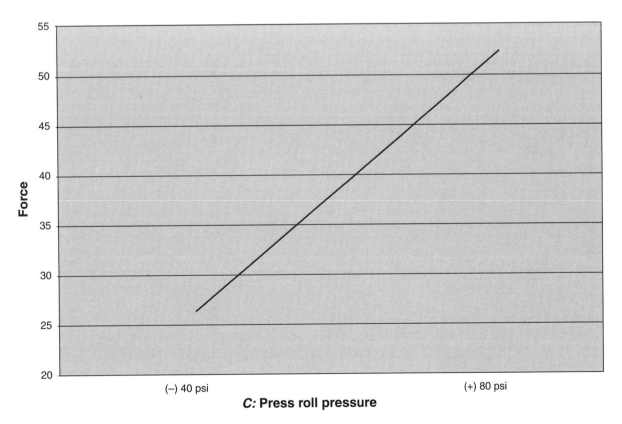

FIGURE 4.3 Effect of C: Press roll pressure.

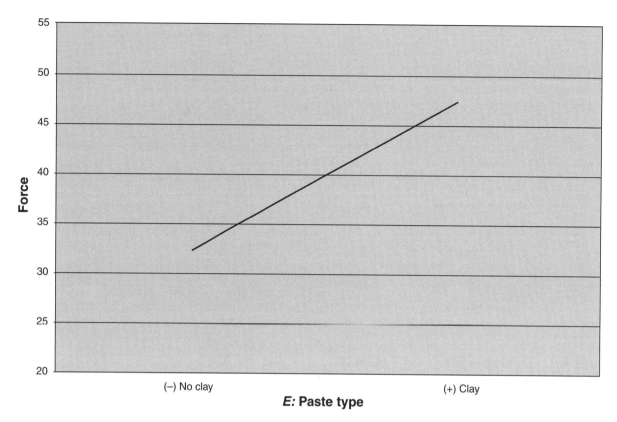

FIGURE 4.4 Effect of E: Paste type.

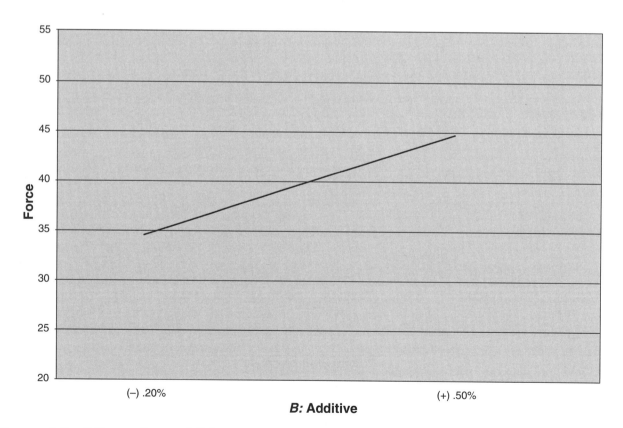

FIGURE 4.5 Effect of B: Additive.

8. Model and conclusions.

The plot of the effects shows that C (press roll pressure), E (paste type), and B (additive) are significant. Since the objective was to maximize the force required to puncture the paperboard, reference to the individual plots shows that this is achieved when these significant factors are set to their high (+) levels. The model for effects is

$$\hat{Y} = \bar{\bar{Y}} + \{E(C)/2\}C + \{E(E)/2\}E + \{E(B)/2\}B =$$
$$39.68 + 12.45C + 7.39E + 5.20B$$

If one substitutes the code for the plus levels, the maximum is predicted to be

$$Y = 39.68 + 12.45(1) + 7.39(1) + 5.20(1) = 64.72$$

Now check the results of the F–tests. Using a test criteria for a risk of 10 percent and degrees of freedom of (4,4),

$$F_{4,4} = 6.39$$

Based on this criterion, only factor F (cure time) influenced the variation in the process. Reference to the analysis table indicates that the low condition had a variance of $S_-^2 = 2.94$ or $S_- = 1.71$. Using the high level of factor F would result in a standard deviation of $S_+ = \sqrt{22.18} = 4.71$, a substantial increase in process variability! What is the *low* condition? Reference to the variable and level list shows that this is the condition with cure time of ten days, not five days! This is to make a point: most people put quantitative levels in numerical order. This is not necessary as long as one uses consistency in labeling. As a famous old baseball umpire once said about a pitch, "It ain't nothing until I call it!" And the same is true of the labeling of levels.

The conclusions are that factors C, B and E can control the process average while factor F can reduce spread or variation. The other factors can be set at convenient or economic levels since they do not appear to influence either the process average or the process spread. What concerns are there for the confounding risks? The design table provided the aliases. Table 4.7 is a recap of the aliases of the three significant factors.

TABLE 4.7. Aliases of the significant factors.

C	E	B
–AG	–AF	–AD
–BE	–BC	–CE
–DF	–DG	–FG

This table indicates that what one thinks is a main effect could possibly be an interaction. Before panicking, refer to the *two critical characteristics* (heredity and sparsity). These rules, *while not guarantees*, state that experience has shown that large main effects often have interactions *and* that main effects usually are much larger than interactions. In Table 4.7, there is only one interaction confounded with each main effect that meets the heredity rule. For example, it is unlikely that AG or DF interactions are real since their main effects showed no indication of being significant. Then, the confounding pattern of C = –BE is the only pattern to worry about for factor C. Similarly, E = –BC and B = –CE are also concerns. Knowledge of the process is also quite useful in ruling out possible interactions. The production people do not see a logical reason for an interaction of *BC* (additive in paper stock and press roll pressure) or *BE* (additive and paste type). *CE* (press roll pressure and paste type) is a logical possibility. However, if the experimenter was confident that the answers to these questions were known, he or she would not be performing this experiment nor would he or she have a problem with this process! The time has come to describe the insurance policy that can factually answer this question or at least assure that the conclusions on main effects are clear.

Reflection

A *reflection* (called a *fold-over* design in some references) is the exact opposite of the basic design in signs. It is achieved by replacing all plus signs with minus signs, and vice versa. The reflected design reverses the signs of the two-factor interactions in the confounding pattern. Let me describe the meaning of the confounding pattern more precisely using only factor *A* of the current design. When it is stated that the confounding is such that $A = -BD = -CG = -EF$, this really means that the estimated effect of *A* is a combination of the actual values of the effects of *A*, *BD*, *CG*, and *EF* in the relationship shown in the base design:

$$E(A)_{est} = E(A)_{act} - E(BD)_{act} - E(CG)_{act} - E(EF)_{act}$$

The reflection provides a confounding pattern opposite in sign such that the reflected design is

$$E(A)_{est} = E(A)_{act} + E(BD)_{act} + E(CG)_{act} + E(EF)_{act}$$

If the results of a reflected design are added to that of a base design and the average calculated, the two equations show that the interaction terms cancel, leaving an estimate of main effects free from confounded two-factor interactions:

$$\tfrac{1}{2}\{\text{base } E(A)_{est} + \text{reflected } E(A)_{est}\} =$$
$$\tfrac{1}{2}\{[E(A)_{est} - E(BD)_{act} - E(CG)_{act} - E(EF)_{act}] + [E(A)_{act} + E(BD)_{act} + E(CG)_{act} + E(EF)_{act}]\} = \tfrac{1}{2}\{E(A)_{act} + E(A)_{act}\} = E(A)_{act}$$

This is the characteristic and purpose of reflection in screening designs: *combining a reflection with a base design will* always *provide estimates of main effects that are free of two-factor interactions.* The two-factor interactions will still be confounded with one another. In the current example, to determine whether the three main effects are realistic or whether some main effect–interaction combination is the actual model, one needs only to run a reflection.

Before proceeding, there is one other aspect of the foregoing relationships that can be used to provide information. If averaging a reflection with the base design clarifies main effects, what if one took the average differences of the effects, i.e., $\{\text{reflected } E(A)_{est} - \text{base } E(A)_{est}\}/2$? These relationships show that such a calculation estimates the combined effects of the confounded interactions:

$$\tfrac{1}{2}[\text{reflected } E(A)_{est} - \text{base } E(A)_{est}\} =$$
$$\tfrac{1}{2}\{[E(A)_{act} + E(BD)_{act} + E(CG)_{act} + E(EF)_{act}] - [E(A)_{act} - E(BD)_{act} - E(CG)_{act} - E(EF)_{act}]\} = \tfrac{1}{2}\{2\,E(BD)_{act} + 2\,E(CG)_{act} + 2\,E(EF)_{act}\} = E(BD)_{act} + E(CG)_{act} + E(EF)_{act}$$

While the individual interactions cannot be identified, this calculation at least gives some indication of the existence of interactions.

This is more easily demonstrated than described by analyzing a reflection of the current design and reanalyzing the results. (This does not have to be done in two steps; one could have planned this from the start and done only one analysis of the sixteen runs.) Table 4.8 shows the reflected design and analysis, including the aliases (notice the reversed signs in the design matrix and for the confounded interactions).

TABLE 4.8. Reflection of Example 4.1.

	A	B	C	D	E	F	G	$\bar{Y}$	S^2
1	−	+	+	−	+	−	−	55.27	6.44
2	−	−	+	+	−	+	−	44.79	9.25
3	−	−	−	+	+	−	+	41.00	2.58
4	+	−	−	−	+	+	−	41.33	11.71
5	−	+	−	−	−	+	+	22.59	21.39
6	+	−	+	−	−	−	+	43.66	2.86
7	+	+	−	+	−	−	−	22.74	1.92
8	+	+	+	+	+	+	+	59.45	27.38
ΣY_+	167.18	160.05	203.17	167.98	197.05	168.16	166.70		
ΣY_-	163.65	170.78	127.66	162.85	133.78	162.67	164.13		
$\bar{Y}_+$	41.80	40.01	50.79	42.0	49.26	42.04	41.68		
$\bar{Y}_-$	40.91	42.70	31.92	40.71	33.45	40.67	41.03		
Effect	.89	−2.69	18.87	1.29	15.81	1.37	.65		
$\bar{S}_+^2$	10.97	12.99	10.19	10.28	10.73	17.43	13.55		
$\bar{S}_-^2$	8.62	6.60	9.40	9.31	8.86	2.16	6.04		
F	1.27	1.97	1.08	1.10	1.21	8.07	2.24		
Aliases	BD	AD	AG	AB	AF	AE	AC		
	CG	CE	BE	CF	BC	BG	BF		
	EF	FG	DF	EG	DG	CD	DE		

As stated earlier, the average of the two sets of effects (base and reflected) provides estimates of main effects that are free of any two-factor interaction effects (see Table 4.9).

TABLE 4.9. Summary of effects.

Factors	A	B	C	D	E	F	G
Base Design (b)	1.13	10.39	24.89	−1.32	14.78	.29	−.06
Reflection (r)	.89	−2.69	18.87	1.29	15.81	1.37	.65
Average = (b+r)/2	1.01	3.85	21.88	−.02	15.30	.83	.30
Difference = (r−b)/2	−.12	−6.54	−3.01	1.31	.52	.54	.36

The averages of the combined estimates provide clean estimates of main effects, and the conclusion is that the effects of C and E remain significant and consistent while the effect of B cannot be distinguished from the inherent variation. Factor B apparently indicates a confounded interaction. The average difference row confirms this, as already discussed, by revealing that the group of interactions confounded with B is significant. Probably only one of the interactions is significant. This group is AD, CE, and FG. The heredity rule from the two critical characteristics of screening designs states that significant main effects often have significant interactions. The most likely candidate is CE since both C and E stand confirmed as significant. This should be verified in a subsequent refining experiment in which the interaction could be assessed directly. Assuming the assessment of CE is correct, can the experimenter get an idea of what this interaction looks like? To do so, condense the entire experiment (base and reflected) into a 2 × 2 table of averages for C and E, using the signs for C and E to place the 16 averages into the appropriate cells (see Table 4.10).

TABLE 4.10. Interaction table for CE.

		C: Press Roll Pressure	
		(−)	(+)
E: Paste Type	(−)	14.67 22.59 14.63 22.74 $\bar{Y} = 18.66$	51.33 44.79 48.53 43.66 $\bar{Y} = 47.08$
	(+)	40.87 41.00 38.77 41.33 $\bar{Y} = 40.49$	54.12 55.27 54.53 59.45 $\bar{Y} = 55.84$

The effect of the interaction can be calculated by the usual formula (it works even for large designs).

$$E(CE) = \tfrac{1}{2}[E(C)_{E+} - E(C)_{E-}] = \tfrac{1}{2}[(Y_{C+} - Y_{C-})_{E+} - (Y_{C+} - Y_{C-})_{E-}]$$
$$= \tfrac{1}{2}[(55.84 - 40.49) - (47.08 - 18.66)] = \tfrac{1}{2}[15.35 - 28.42] = -6.54$$

A CE column could also be created and the effect estimated from that. However, the effect in the E(B) column in the difference row of Table 4.9 *is* the effect of CE plus AD plus FG, and it has been assumed that effects of AD and FG are negligible. Then, E(CE) = −6.54. Since the effect is negative, it reduces some of the benefit anticipated from C and E. Had it been positive, it would have given even more benefit than expected from C and E. Its meaning is best interpreted by a plot of the means, as illustrated in Figure 4.6.

The completed model is

$$\hat{Y} = \bar{\bar{Y}} + \{E(C)/2\}C + \{E(E)/2\}E + \{E(CE)/2\}CE$$

where $\bar{\bar{Y}}$ is the grand average.

$$\hat{Y} = 40.52 + (21.88/2)C + (15.30/2)E - (6.54/2)CE = 40.52 + 10.94C + 7.65E - 3.27CE$$

If C and E are both set at their high levels, each coded as +1, the predicted maximum is

$$\hat{Y} = 40.52 + 10.94C + 7.65E - 3.27CE$$
$$= 40.52 - 10.94(1) + 7.65(1) - 3.27(1)(1) = 55.84$$

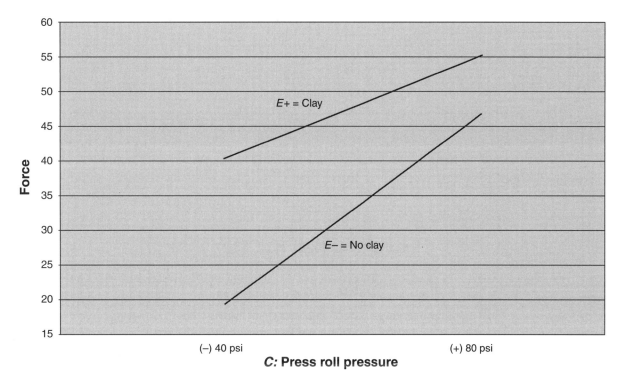

FIGURE 4.6 *CE* Interaction

This is lower than the initial prediction (64.72) using the base design alone. The additional information has modified the model to more accurately reflect (no pun intended!) the interaction. Based on the significant effects, the conclusions now are to use the higher press roll pressure (C_+) and the paste with clay (E_+) to achieve the maximum response. Based on the variation analysis, one would recommend using the longer cure time (F_-) to achieve these results with minimum inherent variation. No other factors appear to have influence beyond that of random variation, and they can be set based on convenience or economics. By the way, did the experimenter check the reflection to see if the influence of F (cure time) on variation was verified? (It was.) How would the refining experiment be run? It would probably be a 2^3 factorial design using C, E, and F. The levels of C (press roll pressure) could be selected not only to verify the relationship but to see if even higher pressure would provide further increase in the response. For example, the pressure might be set at 80 pounds and 100 pounds, instead of the 40 and 80 of the original experiment. One also wishes to see if differences due to paste type disappear at higher roll pressure. (If so, the process would be robust or insensitive to paste type.) Similarly, cure time (F) might be set at 10 days and 8 days instead of the original 10 days and 5 days. Since cure time adds cost, the experimenter is interested in whether it can be reduced without giving up the reduced inherent variation.

Other Analytical Considerations

Replications are expensive. By the time this eight-run Plackett-Burman design was reflected and replicated, 32 trials were required. What if one could not afford to replicate? Could the DOE still be analyzed? Yes! Simply working from the Pareto analysis would have been very effective. Tests of significance are nice but by no means essential. In general, if one has a choice of reflection or replication, reflection should be used. The exception to this rule arises when there is reason to believe that the experiment will be subject to special cause variation or outliers. For a variation analysis, the technique introduced earlier using residuals can be applied when there is no replication.

50   AN INTRODUCTION TO DESIGN OF EXPERIMENTS

In review, the basic screening designs are highly efficient for sifting many factors to select a smaller subset for more detailed study. Their penalty is the confounding of main effects with two-factor interactions—or at least the confounding of interactions with one another. By doubling the number of trials in the experiment, reflection assures clean assessment of main effects with limited information on interactions. Screening designs are a logical launch point for a sequence of experiments that may culminate in optimization studies of a very few factors. In addition, note how easily one can identify an even larger group of factors that are *not* worth further study. This allows the experimenters to address situations where money is being spent to control factors that do not matter!

Exercise 4 should now be worked by the reader. Check the analysis in the Appendix *after* first trying it!

Exercise 4: Nail Pull Study

A team is trying to improve the strength characteristics of a wall board as measured by the force required to pull a nail through the material. The board is made up of layers of paper pasted together and pressed. The final list of factors is listed following showing the low versus high levels. The objective is to maximize the response. There were two replicates of each run, each being a completely random and independent setup.

A: Paste temperature—120°F vs. 150°F

B: Roll pressure—40 psi vs. 80 psi

C: Amount of sizing—.5 percent vs. 1.0 percent

D: Paste types—X vs. Y

E: Paper moisture—4 percent vs. 8 percent

F: Cure time—5 days vs. 10 days

G: Mill speed—200 fpm vs. 250 fpm

The factors vary in their expense to the process:

1. Extending cure time creates a massive and expensive storage problem.
2. Mill speed is directly correlated with productivity.
3. High-moisture paper is cheaper than low-moisture paper.
4. Increasing the amount of sizing adds to the cost of board.
5. Temperature and paste type are very minor in their cost impact.
6. Roll pressure has no cost impact.

Analyze and make recommendations to management. The analysis table is shown in Table 4.11. Upon completing your work, refer to the results in the Appendix. Complete the missing columns as part of the exercise.

TABLE 4.11. Analysis table for Exercise 4.

	A	B	C	D	E	F	G	$\bar{Y}$	S^2
1	+	−	−	+	−	+	+	59.75	.61
2	+	+	−	−	+	−	+	56.20	38.69
3	+	+	+	−	−	+	−	70.15	3.65
4	−	+	+	+	−	−	+	72.60	.02
5	+	−	+	+	+	−	−	70.90	7.24
6	−	+	−	+	+	+	−	55.55	73.27
7	−	−	+	−	+	+	+	63.20	42.10
8	−	−	−	−	−	−	−	50.00	12.50
ΣY_+	257.00		276.85	258.80		248.65	251.75		
ΣY_-	241.35		221.50	234.55		249.70	246.60		
$\bar{Y}_+$	64.25		69.21	64.70		62.16	62.94		
$\bar{Y}_-$	60.34		55.37	59.89		62.43	61.65		
Effect	3.91		13.84	4.81		−0.26	1.29		
$\bar{S}_+^2$	12.55		13.25	20.29		29.91	20.36		
$\bar{S}_-^2$	31.97		31.27	24.24		14.61	24.17		
F	2.55		2.36	1.19		2.05	1.19		
Aliases	−BD	−AD	−AG	−AB	−AF	−AE	−AC		
	−CG	−CE	−BE	−CF	−BC	−BG	−BF		
	−EF	−FG	−DF	−EG	−DG	−CD	−DE		

Twelve-Run Plackett-Burman

The next design to be demonstrated is a 12-run Plackett-Burman that allows up to 11 factors in 12 trials. Since this is a nongeometric design, the main effects are partially confounded with all two-factor interactions that do not contain that main effect. For example, the main effect A is partially confounded with all 45 of the two-factor interactions that do not contain A! While this sounds serious, be aware that so many pieces of interactions have a good chance of averaging out to a very small effect. In addition, if an interaction is significant, only a fraction of its effect contributes to the confounding of the main effects. An example will help clarify these issues.

Example 5: Moldability Analysis

The following experiment studied the moldability of an automotive door part. The response was the angle of radius. The part must be flexible and *angle of radius* refers to how far the part can be bent before fracturing. Brainstorming resulted in about 25 factors initially. Subsequent paring based on cost, control, and time issues (look initially for factors that are quick, cheap, and easy) produced 11 factors for this initial screening design. It was considered too expensive to replicate or reflect the design. The objective is to maximize the angle of radius or bend. The analysis matrix is shown in Table 4.12.

TABLE 4.12. Analysis matrix for Example 5.

	A	B	C	D	E	F	G	H	I	J	K	Y
1	+	−	+	−	−	−	+	+	+	−	+	59.48
2	+	+	−	+	−	−	−	+	+	+	−	73.90
3	−	+	+	−	+	−	−	−	+	+	+	57.96
4	+	−	+	+	−	+	−	−	−	+	+	56.37
5	+	+	−	+	+	−	+	−	−	−	+	71.49
6	+	+	+	−	+	+	−	+	−	−	−	42.69
7	−	+	+	+	−	+	+	−	+	−	−	60.57
8	−	−	+	+	+	−	+	+	−	+	−	70.10
9	−	−	−	+	+	+	−	+	+	−	+	56.20
10	+	−	−	−	+	+	+	−	+	+	−	42.58
11	−	+	−	−	−	+	+	+	−	+	+	48.07
12	−	−	−	−	−	−	−	−	−	−	−	54.30
ΣY₊	346.51	354.68	347.17	388.63	341.02	306.48	352.29	350.44	350.69	348.98	349.57	
ΣY₋	347.20	339.03	346.54	305.08	352.69	387.23	331.42	343.27	343.02	344.73	344.14	
Ȳ₊	57.75	59.11	57.86	64.77	56.84	51.08	58.72	58.41	58.45	58.16	58.26	
Ȳ₋	57.87	56.51	57.76	50.85	58.78	64.54	56.90	57.21	57.17	57.46	57.36	
Effect	−.12	2.60	.10	13.92	−1.94	−13.46	1.82	1.20	1.28	.71	.90	

Since there is no replication, analysis must be based on the Pareto of effects (see Figure 4.7).

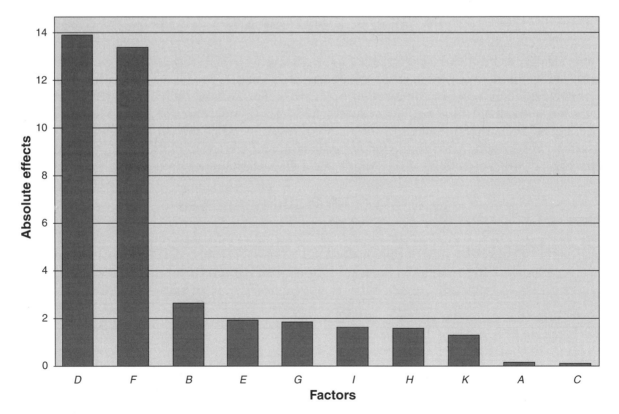

FIGURE 4.7. Pareto of effects for Example 5.

The *eyeball* assessment would appear rather conclusive: only factors D and F are dominant. At this point, the usual plots of the effects for D and F are made to make interpretation easier (see Figures 4.8 and 4.9).

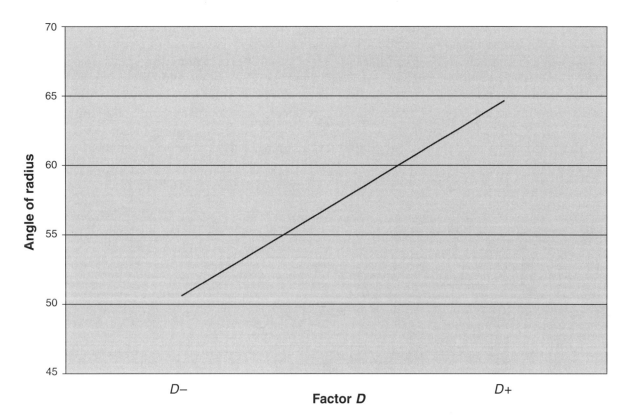

FIGURE 4.8. Effect of D.

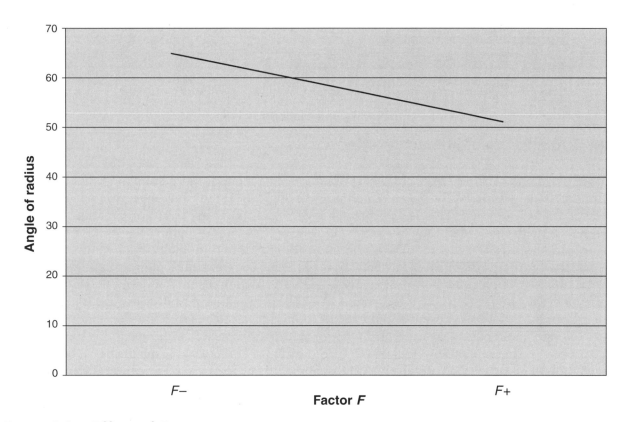

FIGURE 4.9. Effect of F.

Recall the previous concerns about the potentially confusing impact of interactions. With respect to this concern, one can again take great comfort in the *two characteristics of screening design*. The sparsity rule states that it is rare to find an interaction dominating main effects, and the heredity rule suggests that large main effects often have interactions. One can be confident that the two main effects here are indeed valid. To confirm this belief, the design may be reflected to remove the confounding influences of the interactions or a small factorial may be used to validate and refine the conclusions. But can one detect interactions directly? If there are only a couple of dominating main effects, it is possible to do limited checking for interactions. Since there are 55 two-factor interactions, one obviously can't evaluate them all! However, the heredity rule provides a guide for a limited assessment: check the interactions involving D or F. The first choice is to check for a DF interaction. Following that, check the remaining 18 interactions involving D or F. (This can be done by creating interaction columns. A spread sheet should be used for this.) Figure 4.10 is a Pareto of those interaction effects.

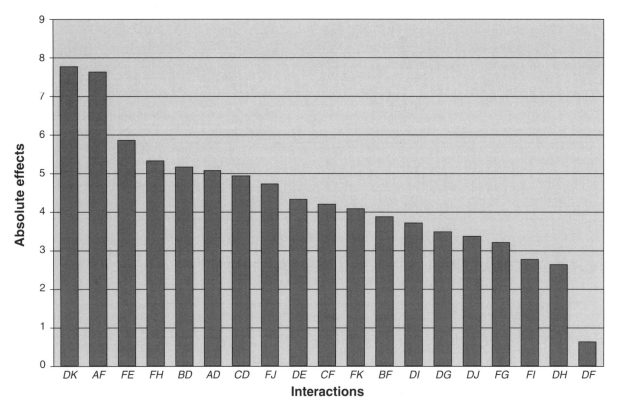

FIGURE 4.10. Pareto of interaction effects.

What does one conclude from this chart of effects. Realize that there are either a lot of significant interactions—or probably none. Which is more logical? While they exist, highly interactive processes are not common. That is, few processes are dominated by several interactions. If this *is* one, a screening design is definitely the wrong tool since screening designs focus on identifying factors for further study and are not focused on interactions. More than likely, there is no interaction here, but it would take the next experiment to verify this. The next experiment might include factors D and F plus any factors that were involved with a questionable interaction; e.g., since DK and AF are the largest interactions, factors K and A might be added to the next factor list. If at all possible, a reflection of the base design should be performed to assure correct interpretations of main effects. (Nongeometric designs are weaker than geometric designs in handling either a highly interactive process or that rare process that is dominated by an interaction. This results from the fact that significant interactions in nongeometric designs contribute to many main effects and may provide misleading estimates of effects. If the process is dominated by an interaction, there may be no hints of such a situation until a verification study.)

Based on these results, the model is

$$\hat{\hat{Y}} = \bar{\bar{Y}} + \{E(D)/2\}D + \{E(F)/2\}F$$

From table 4.13,

$$\hat{\hat{Y}} = 57.81 + (13.92/2)D + (-13.46/2)F$$

and

$$\hat{\hat{Y}} = 57.81 + 6.96D - 6.73F$$

Since the objective is to maximize the response, this occurs when D is at the high setting and F is at the low setting. For this combination, the predicted response is 71.50. (A quick verification check would be to operate at this condition and see if the results are close to this value.) Has the process been improved? That depends on what the normal operating conditions have been. If these (D_+ and F_-) were the typical conditions, the experimenter has not improved the process but only verified the importance of continuing to control these factors. If these are not the typical operating conditions, this model indicates that an improvement is possible. The size of the improvement depends on the comparison of this forecast and the average of typical performance.

The primary objective of this study has now been accomplished: screen eleven factors and select those that appear worthy of a more detailed investigation. If one has missed an interaction (or even a smaller main effect), he or she still has identified some of the major factors in the process and defined subsequent refining experiments to confirm and explore the relationships further. The primary characteristics of any nongeometric design is both a strength and a weakness: partial interactions are confounded everywhere. This gives a good chance that they will balance out (some will be positive and some will be negative) and permits direct estimation of one or two interactons. The experimenter can also be unlucky and be misled in identifying significant main effects. If the calculation of multiple interactions becomes a key part of the analysis, this is the wrong design and reflection is essential to provide unbiased estimates of the main effects.

Exercise 5 follows for the reader to practice what has been covered. The results are in the Appendix for review after the analysis has been completed.

Exercise 5: The Lawn Fanatics

A lawn service wanted to analyze the factors that lead to a high-growth lawn. They had a very homogeneous five-acre plot that was used for the experiment. They determined the following ten factors to investigate:

- A: Amount of water per week—1" vs. 2"
- B: Amount of lime—2# per plot vs. 5#
- C: Amount of nitrogen—.10 percent vs. .30 percent
- D: Amount of phosphorous—.1 percent vs. .2 percent
- E: Grass type—Brand X vs. Brand Y
- F: Blade condition—dull vs. sharp
- G: Grass height after mowing—2 inches vs. 2.5 inches
- H: Insecticide—none vs. regular application
- I: Weed treatment—none vs. regular application
- J: Time of day of watering—5 A.M. vs. 4 P.M.

The design was a 12-run Plackett-Burman with reflection. The field was divided into 24 plots, each two-tenths of an acre. The 24 trials were assigned to the plots by drawing numbers from a box. The response was the average weight of the grass clippings from the plots, averaged over six mowings. The mowings were seven days apart.

The analysis table is shown in Table 4.13. Analyze, generate conclusions and recommendations, and critique the experiment.

TABLE 4.13. Analysis table for Exercise 5. (12-run Plackett-Burman-Reflected)

	A	B	C	D	E	F	G	H	I	J	K	Y
1	+	−	+	−	−	−	+	+	+	−	+	53.0
2	+	+	−	+	−	−	−	+	+	+	−	63.2
3	−	+	+	−	+	−	−	−	+	+	+	48.2
4	+	−	+	+	−	+	−	−	−	+	+	58.7
5	+	+	−	+	+	−	+	−	−	−	+	53.4
6	+	+	+	−	+	+	−	+	−	−	−	71.6
7	−	+	+	+	−	+	+	−	+	−	−	50.5
8	−	−	+	+	+	−	+	+	−	+	−	44.3
9	−	−	−	+	+	+	−	+	+	−	+	31.2
10	+	−	−	−	+	+	+	−	+	+	−	48.4
11	−	+	−	−	−	+	+	+	−	+	+	40.7
12	−	−	−	−	−	−	−	−	−	−	−	25.0
13	−	+	−	+	+	+	−	−	−	+	−	45.4
14	−	−	+	−	+	+	+	−	−	−	+	46.4
15	+	−	−	+	−	+	+	+	−	−	−	44.2
16	−	+	−	−	+	−	+	+	+	−	−	47.1
17	−	−	+	−	−	+	−	+	+	+	−	43.2
18	−	−	−	+	−	−	+	−	+	+	+	30.9
19	+	−	−	−	+	−	−	+	−	+	+	48.0
20	+	+	−	−	−	+	−	−	+	−	+	54.8
21	+	+	+	−	−	−	+	−	−	+	−	76.0
22	−	+	+	+	−	−	−	+	−	−	+	55.7
23	+	−	+	+	+	−	−	−	+	−	−	60.3
24	+	+	+	+	+	+	+	+	+	+	+	65.0
ΣY_+	696.6	671.6	672.9	602.8	609.3	600.1	599.9	607.2	595.8	612.0	586.0	
ΣY_-	508.6	533.6	532.3	602.4	595.9	605.1	605.3	598.0	609.4	593.2	619.2	
Y_+	58.05	55.97	56.08	50.23	50.78	50.01	49.99	50.60	49.65	51.00	48.83	
Y_-	42.38	44.47	44.36	50.20	49.66	50.43	50.44	49.83	50.78	49.43	51.60	
Effect	15.67	11.50	11.72	.03	1.12	−.42	−.45	.77	−1.13	1.57	−2.77	

Even Larger Designs

These principles of design and analysis are used with even larger designs. The 16-run, in particular, is a major workhorse. The size of the experiment is limited by the ability of the experimenter to afford and control very large experiments. From a mathematical basis, larger is better since there are more terms that tend to average the impact of confounding to a net zero. From the standpoint of the experimenter who is trying to assure that the correct treatment combinations are performed, there is a point at which control becomes difficult. Cost and control are the realistic limits to the size of an experiment. The Appendix contains designs up to 20-run Placket-Burman and provides analysis tables for each. Software will provide even larger designs and provide a choice of Plackett-Burman or fractional factorial designs.

Other Types of Designs

This has been an introduction to the basic designs—those with two levels. Three-level designs can also be generated as either full factorials or as fractional factorials when nonlinearity is a major concern. The analysis of such designs is not as easy as the two-level designs, and more runs are necessary. There are also some unusual Taguchi designs that combine two-level and three-level factors. One of the more useful of these is the L-18 design, which will analyze one factor at two levels and seven factors at three levels, requiring only 18 runs. Additionally, there are highly efficient designs called *response surface* designs that expand small two-level designs to efficiently address curvature. These are often used to model processes in the final refining or optimization stage. (The subject of optimization designs definitely merits further study by the reader.) *Evolutionary Operation* (EVOP) is a technique for optimizing a process while it is operating. A very different family of designs is the family of *mixture designs,* which are appropriate when the factors' levels must total to some fixed percentage of the total contents of a material. These and other designs are discussed in advanced DOE training and must be presented with software to facilitate their analysis.

At this point, the reader must again be reminded that this has been an introduction of DOE. The intent has been to get you into *doing* experiments and to encourage you to go on to study more advanced references. Many valid points have been skipped or glossed over in the interest of simplicity. The reader is strongly advised to review any of the reference texts listed at the end of Chapter 5 to gain exposure to many of the more advanced techniques and concepts.

Problems, Questions, and Review

Problems and Questions

Frequently asked questions and technical points that need to be repeated follow:

1. *If one cannot do both, should the experimenter reflect or replicate a screening design?*

   Reflection is always preferred *except* when there is concern about special cause variation during the DOE. In such cases, replication is preferred. The rationale is that assessment of significance can be done quite well with a Pareto chart, partially overcoming the lack of a formal test of significance that is possible with replication. Reflection provides clarity of results on main effects. However, if special causes abound, one will have difficulty drawing accurate conclusions without replication. Assessment of interaction is very risky in the face of random special causes.

2. *What if many of the effects are significant?*

   The experimenter should carefully review the randomization pattern used while performing the experiment. Was the setup for *each* trial really randomly replicated? If not, the responses are *repeats*, not *replicates*. *Replication* assumes that each trial is an independent and random performance of the process, specifically including any process setup. If not, the experimenter has *repeats* that may be neither independent nor random. The estimate of S_e using repeats will be much smaller than actual inherent variation of the process. This will make most effects appear significant, when in reality, they have not been randomly replicated to properly estimate the inherent variation.

3. *What if several factors are significant in the variation analysis?*

   Variation analysis is very susceptible to outliers or special causes that create major distortions of the variances of each run. Check to see if one or two estimated variances are substantially larger than the others. Review the raw data for outliers, i.e., unusual values compared to the other observations. Apparent outliers should be investigated and, if determined to be outliers, removed or replaced with the cell average.

4. *Can one analyze an experiment with missing data?*

   The simplest thing to do is to replace a missing observation with the grand average if the experiment is unreplicated or with the average of the other observations for that run if replicated. This minimizes the loss of information. If more than one observation is lost in an unreplicated DOE, computer analysis may be the only way to salvage the analysis. (Note that this is a nonissue when one moves to software support.)

5. *Does one have to use the maximum number of factors for a design?*

   No. While all *runs* of a design must be carried out, one may use less than the maximum number of factors. For example, in a 16-run Plackett-Burman design, there is nothing wrong with having fewer than fifteen factors. The unused factors are often referred to as *dummy factors*. The columns of the unused factors represent the family of interactions that would ordinarily have been confounded with that factor(s) had it been used. Interactions with the unused factor(s) do not exist.

   It is to the experimenter's advantage to study as many factors as possible in an experiment. Don't use fewer than the maximum factors of a design if there are any reasonable factors that could be added to the factor list.

6. *Should one use Plackett-Burman designs or fractional factorial designs?*

   This question must be answered in two parts.

   a. Geometric designs, whether Plackett-Burman or fractional factorials, are equivalent as long as the maximum number of factors is used in Plackett-Burman designs. For example, an eight-run Plackett-Burman with seven factors is equivalent to an eight-run fractional factorial design with seven factors. However, as fewer than seven factors are used, there comes a point where fractional factorials provide better optimization of the confounding of two-factor interactions. That is why eight-run fractional factorial designs are provided in the appendix for four and five factors. This same point holds for larger geometric designs, e.g., the 16-run or 32-run designs. The fewer the factors (below the maximum possible), the more advantageous the fractional factorial designs will be in minimizing confounding.

   b. Nongeometric designs are uniquely Plackett-Burman designs with the disadvantages discussed in Chapter 4. Nongeometric designs incur more risks than geometric designs in accurately estimating main effects. However, the experimenter should use a nongeometric design when it fits his or her needs. For instance, if there are ten or eleven factors to be studied, one should not use a 16-run geometric design just to avoid a 12-run nongeometric design. There is, as usual, an exception: if the process is known or suspected to be highly interactive, it would be better to have a geometric design in order to provide at least a grouping of interactions.

   Note that reflection is essential to provide clean estimates of main effects in either case! Screening designs are not meant to provide major information on interactions.

7. *What is* blocking *and how is it used?*

   There are times when the experiment cannot be completely randomized. For instance, suppose there is not enough raw material to complete the experiment with one lot and the experimenter fears that lot differences could create a distortion in the responses. This could be handled by letting *raw material lot* be a factor and randomizing *within* each lot. Another example would be an experiment that could not be completed in one day, and a time effect was considered a possibility. Again, let one factor be *day* with day one being the high level and day two being the low level; randomize all the trials within each day.

8. *What if one can replicate a few runs but cannot afford to replicate all of them?*

   a. Randomly select the runs to be replicated and add these trials to the design.

   b. After the data are available, calculate variances for each replicated run.

   c. Calculate the average variance; this is the estimate of S_e^2. The degrees of freedom are

   $$(\text{no. of replicates} - 1) \times (\text{no. of replicated runs})$$

d. Use the average of the replicates in the analysis table and proceed as usual to estimate effects.

e. Use the calculated S_e^2 for significance testing. Do not perform the variation analysis.

9. *Should one run one big screening experiment and quit?*

   No! Experimentation is usually a series of experiments, each building from the other. The experiments are generally of diminishing size as the studies move from screening to optimization. While we all get lucky sometimes, it is not typical to get all of your answers in the first experiment.

10. *Must one randomize all trials?*

    No, although it would be the safest strategy. If there is concern about the feasibility of a run, it would be prudent to carry it out first in case the levels must be redefined. It is best to learn this early before the study has really gotten under way.

    In other experiments, a physical constraint may prevent randomization. Assume a furnace temperature must be set and then trials run for that temperature. (In other words, one can't bounce back and forth.) Be aware that this flinching on the randomization could lead to confounding temperature with an unknown blocking factor.

There are many other situations that may arise. The main rules to remember in dealing with peculiar results are

- Use common sense and process knowledge.
- Look at the data!

Review of the Basics in Managing a DOE

Most people assume that the analytical techniques are the most difficult part of applying DOE. This is not true. The most difficult aspects of any DOE are the planning of the experiment and the actual running of the experiment. Following is a review of the basics of application of DOE:

1. Identify the process to be studied.
2. Precisely define the objectives of the study.
3. Assemble a small team knowledgeable in the process being addressed.
4. Review all pertinent data and knowledge relevant to the process.
5. Brainstorm to generate potential factors. Be creative, and do not accept existing theories without data.
6. Distill from this list those factors that can be controlled and whose cost to the experiment can be tolerated. (The more, the better.) Be sure to include those factors that are cheap, quick, and easy to study.
7. Set levels boldly without being careless. One needs levels as wide as reasonable to force effects to show themselves. Yet, process knowledge must be used to avoid conditions that are either not feasible or dangerous.
8. Select the design that best fits your study. Determine the number of factors to be studied; review the cost and complexity of control of the experiment; consider the need for reflection and for replication.
9. Randomize the order of the DOE as best you can without severe penalty. (And, if you cannot randomize completely, make sure that you are aware of and sensitive to the potential impact of this lack of randomization.) This protects against drawing false conclusions because some

factor outside the experiment influenced the response at the same time that a DOE factor was changed in level. This is the insurance policy for unbiased conclusions.

10. Perform the experiment. Make sure levels are correctly set for each trial and that materials are identified. Keep good records on all aspects of the DOE.
11. Report the conclusions and recommendations in the language of the audience, not in the statistical terminology of the statistical specialist. Graphs communicate quite well; use them extensively.

What Inhibits Application of DOE?

Despite its power and versatility, there are several reasons why DOE is not more commonly used. The following is a short list of reasons, not necessarily in order of priority.

1. *Inertia.* "We've always studied one factor at a time and I don't have time to learn a new approach." However, if you always do what you've always done, you will always get what you've always gotten!!
2. *Technical experts.* This hits every one of us. If we are so confident that we already know all the answers, why is this process unsatisfactory? If we are to achieve breakthrough in any technology (manufacturing, marketing, sales, research, etc.), we must be willing to challenge what we believe. Remember that *sometimes we do know the answers; sometimes we believe firmly in myths; and sometimes we don't even know the question!* Have you ever noticed that major breakthroughs were generally ranked low on the initial Pareto chart of potential answers?
3. *Cost.* How often do we spend a fortune on rework, scrap, or poor sales results but can't afford a DOE to factually resolve the problem once and for all? Nothing is free. DOE is an investment with a payback to be measured just like a capital project.
4. *Manpower.* The major cost of a DOE is often not in materials or lost process time but in people's time. This is a resource to be managed and prioritized against a return similar to a capital justification.
5. *Lack of management support.* DOE is a management tool; it cannot be used by line personnel without the concurrence and support of management. Since DOE requires commitment of resources and discipline in performance of the trials, management must be involved. That means management must be educated to the point of understanding what the objectives are, what the costs will be, what should be expected from the study, and what the potential power of DOE is.
6. *Training.* Without a few people educated in the techniques, it is difficult, if not impossible, to have lasting and ongoing impact from DOE.
7. *Assumption that DOE is only a manufacturing tool.* DOE is very useful in determining what actions or strategies are effective in increasing sales or market share or answering any other question where options can be defined as factors for study. Is it more difficult in nonmanufacturing areas? Probably. But the payoff is also greater!

Software

Now that the basics have been covered, it is time to discuss software. With an understanding of the basic techniques, there are several software packages that are helpful in carrying out the detailed calculations. This author strongly advises becoming familiar with one. The following are three available options:

Spreadsheets. All of the calculations performed in this book could be generated very quickly in a spreadsheet. Even the charts can be made with relative ease. Such spreadsheets can be customized to the experimenter's liking.

DesignExpert™. This product is tailored specifically for DOE and is quite powerful, flexible, and easy to use. The manual provides a lot of education on advanced topics. This software also handles response surfaces and mixtures. Available from STAT-EASE, Inc., in Minneapolis, MN (800-801-7191).

Number Cruncher (NCSS)™. This software is a *very* powerful and general statistical analysis package. It is user friendly and very flexible in analyzing difficult or messy data. It covers any number of levels, response surfaces, mixtures, and multiple regression. This is a *complete* statistics package that covers far more than just DOE. It is the most user friendly and economical of the major statistical packages. Available from NCSS in Kaysville, UT (801-546-0445).

There are other software products available; these are simply the ones with which the author is personally familiar.

Appendix

The Appendix contains the necessary statistical tables and the answers to the exercises. In addition, design and analysis tables are provided for designs up to 20 runs. These include with and without reflection.

Conclusion

This has been an introduction to design of experiments, the purpose of which was to make the reader capable of handling 90+ percent of the DOE requirements that he or she may come across. There is much more material to be studied for those interested in more depth. Such things as optimization designs, Taguchi designs, EVOP, and mixture designs all have their place. Several complex issues have been avoided in the interest of simplicity but with little damage to the validity of the techniques that have been presented. However, if one starts experimenting using only what has been covered in this book, he or she will make progress much more rapidly than without it.

There is an overwhelming amount of reference material available. A short list of the author's favorites, which combine depth and clarity in their presentations, is included in the next section. Hopefully, the reader has been persuaded to proceed further into this fascinating field.

The main challenge is to just GO DO IT! Inertia is the bane of progress. And keep your wits about you. Good Luck!!

References

Box, G. E., W. G. Hunter, and J. S. Hunter. 1978. *Statistics for Experimenters.* New York, NY: J. Wiley & Sons.

Montgomery, D. 1991. *Design and Analysis of Experiments.* New York, NY: J. Wiley & Sons.

Schmidt, S. R., and R. G. Launsby. 1992. *Understanding Industrial Designed Experiments.* Colorado Springs, CO: Air Academy Press.

Wheeler, D. J. 1990. *Understanding Industrial Experiments.* Knoxville, TN: SPC Press.

Appendix

Table A.1: Table of t-Values for $\alpha = .05$ (95% confidence)   66

Table A.2: F-Table for $\alpha = .10$ 67

Exercise 1: Coal-Processing Yield   68

Exercise 2: Corrosion Study   72

Exercise 3: Ink Transfer   76

Exercise 4: Nail Pull Study   81

Exercise 5: The Lawn Fanatics   85

Tables A.11 and A.12: Analysis Tables for 2^3 Factorial   90–91

Table A.13: Design Matrix for $\frac{1}{2}$ Fractional Factorial for Four Factors   92

Table A.14: Analysis Table for $\frac{1}{2}$ Fractional Factorial with Four Factors   93

Table A.15: Design Matrix for $\frac{1}{4}$ Fractional Factorial for Five Factors   94

Table A.16: Analysis Table for $\frac{1}{4}$ Fractional Factorial with Five Factors   95

Table A.17: Design Table for $\frac{1}{2}$ Fractional Factorial with Five Factors   96

Table A.18: Analysis table for $\frac{1}{2}$ Fractional Factorial with Five Factors   97

Table A.19: Confounding pattern for Eight-Run Plackett-Burman Design   98

Table A.20: Confounding Pattern for 16-Run Plackett-Burman Design   98

Table A.21: Analysis Table for Eight-Run Plackett-Burman Design   99

Table A.22: Analysis Table for Reflection of Eight-Run Plackett-Burman Design   100

Table A.23: Analysis Table for Eight-Run Plackett-Burman Design with Reflection   101

Tables A.24 and A.25: Analysis Tables for 12-Run Plackett-Burman Design   102, 103

Table A.26: Analysis Table for 12-Run Plackett-Burman Design with Reflection   104

Table A.27: Analysis Table for 16-Run Plackett-Burman Design   105

Table A.28: Analysis Table for 16-Run Plackett-Burman Design with Reflection   106

Table A.29: Design Matrix for 20-Run Plackett-Burman   107

TABLE A.1. Table of t values for $\alpha = .05$ (95% confidence).

Degrees of Freedom	t-Value	Degrees of Freedom	t-Value
1	12.71	21	2.08
2	4.30	22	2.07
3	3.18	23	2.07
4	2.78	24	2.06
5	2.57	25	2.06
6	2.45	26	2.06
7	2.37	27	2.05
8	2.31	28	2.05
9	2.26	29	2.05
10	2.23	30	2.04
11	2.20	40	2.02
12	2.18	50	2.01
13	2.16	60	2.00
14	2.15	120	1.98
15	2.13	∞	1.96
16	2.12		
17	2.11		
18	2.10		
19	2.09		
20	2.09		

TABLE A.2. F-table for $\alpha = .10$.

Degrees of Freedom in Both Numerator and Denominator	F-Value
2	19.00
3	9.28
4	6.39
5	5.05
6	4.28
7	3.79
8	3.44
9	3.18
10	2.98
11	2.82
12	2.69
13	2.58
14	2.48
15	2.40
16	2.33
18	2.22
20	2.12
22	2.05
24	1.98
26	1.93
28	1.88
30	1.84
32	1.80

Table A.2 is a simplified table for the case in which degrees of freedom are evenly divided between numerator and denominator terms. If more general parameters are needed, refer to the tables of a basic statistics text. This table is the value of F for $\alpha = .05$ in the upper tail; total alpha risk is 10%.

Exercise 1: Coal-Processing Yield

In a coal-processing plant, pulverized coal is fed into a flotation cell. There the coal-enriched particles are attached to a frothing reagent and floated over into the next wash cell. Rock or low-grade coal particles sink and are taken out the bottom as waste. The yield is the proportion of the input coal that is floated into the next washing step. In attempting to enrich the yield of the cell, two factors are being considered:

1. Retention time in the cell—45 seconds versus 60 seconds
2. Flotation reagent concentration in the solution—1 percent versus 2 percent

A 2^2 factorial design was used with four replicates of each of the four runs. The data are shown in Table A.3. The objective is to maximize the yield in this processing step.

TABLE A.3. Data for coal yield.

		A: Retention Time	
		45 sec. (−)	60 sec. (+)
B: Flotation Reagent Concentration	1% (−)	(− −) 57.7 61.5 57.9 53.5 $\overline{Y} = 57.65$ $S^2 = 10.70$	(+ −) 64.5 71.8 69.1 73.4 $\overline{Y} = 69.7$ $S^2 = 15.16$
	2% (+)	(− +) 80.1 83.1 85.1 77.9 $\overline{Y} = 81.55$ $S^2 = 10.14$	(+ +) 79.5 70.2 77.2 73.5 $\overline{Y} = 75.10$ $S^2 = 16.78$

This experiment was analyzed using the eight-step procedure for analysis of effects and using the spreadsheet approach.

1. *Calculate the effects by the spreadsheet method* (see Table A.4).

TABLE A.4. Exercise 1 in spreadsheet format.

	A	B	AB	$\overline{Y}$	S^2
1	−	−	+	57.65	10.70
2	+	−	−	69.70	15.16
3	−	+	−	81.55	10.14
4	+	+	+	75.10	16.78
ΣY_+	144.80	156.65	132.75		
ΣY_-	139.20	127.35	151.25		
$\overline{Y}_+$	72.4	78.33	6.38		
$\overline{Y}_-$	69.6	63.68	75.63		
Effect	2.80	14.65	−9.65		

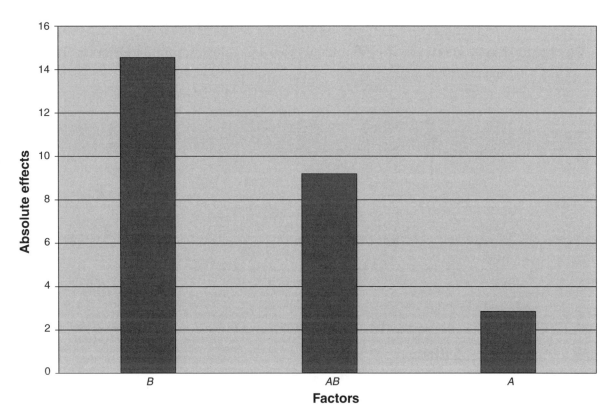

FIGURE A.1. Pareto of effects: Coal processing.

2. *Make a Pareto chart of effects* (see Figure A.1).

3. *Calculate the standard deviation of the experiment, S_e.*

$$S_e = \sqrt{(\Sigma S_i^2 / k)} = \sqrt{[10.70 + 15.16 + 10.14 + 16.78] / 4} = \sqrt{13.195} = 3.63$$

4. *Calculate the standard deviation of the effects.*

$$S_{eff} = S_e \sqrt{4/N} = 3.63\sqrt{4/16} = 1.82$$

5. *Compute degrees of freedom and t.*

df = (number of replicates per -1) × (number of runs) = (4–1) × (4) = 12

For 95 percent confidence and 12 degrees of freedom, t = 2.18.

6. *Decision limits.*

$$DL = \pm(t)\,(S_{eff}) = \pm(2.18)\,(1.82) = \pm 3.97.$$

Any effects outside these limits are significant. See Figure A.2 for graph.

B (reagent concentration) and the AB interaction (time-concentration) are significant.

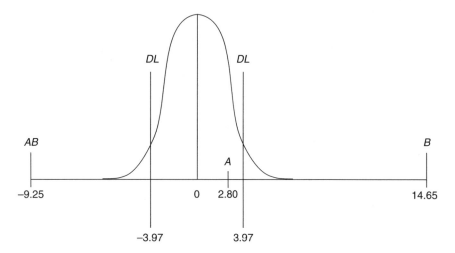

FIGURE A.2. Decision limits.

7. *Plot significant effects* (see Figures A.3 and A.4).

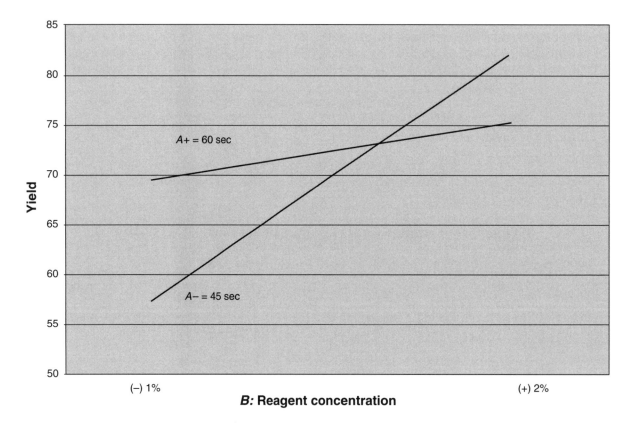

FIGURE A.3. Interaction of *B* (reagent concentration) and *A* (retention time).

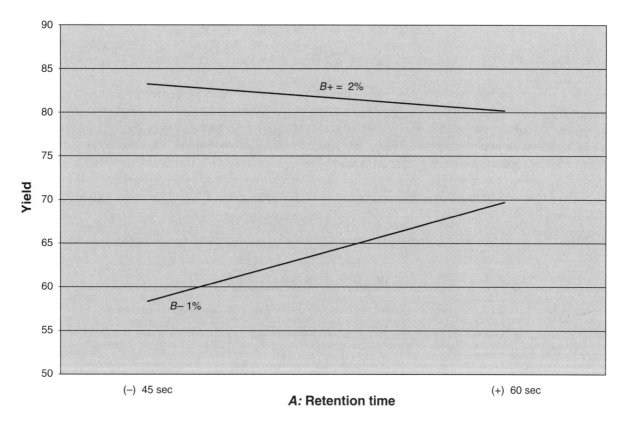

FIGURE A.4. Interaction of *A* (retention time) and *B* (reagent concentration).

Since the interaction is significant, the main effects do not need to be plotted. If in doubt as to which way to show the interaction plot, try it both ways to determine which seems to communicate the best.

8. *Model and Prediction*

Recall that the *hierarchy rule* for modeling requires that all main effects be shown with a significant interaction, even if those terms were not significant. Thus, the model must include a term for factor *A* even though it was not significant.

$$\hat{\bar{Y}} = \bar{\bar{Y}} + [E(A)/2]A + [E(B)/2]B + [E(AB)/2]AB$$
$$= 71.00 + (2.80/2)A + (14.65/2)B - (9.25/2)AB$$
$$= 71.00 + 1.40A + 7.32B - 4.62(AB)$$

From the graphs, it is apparent that the optimum yield would occur when $A- = 45$ sec. and $B+ = 2\%$. At that condition, the model projects a yield of

$$\hat{\bar{Y}} = 71.00 + 1.40(-1) + 7.32(+1) - 4.62(-1)(+1) = 81.54$$

Since lower retention time increases throughput, the next study could address whether a further reduction in retention time could lead to even higher yield with 2 percent reagent concentration. This would be a double benefit. One could also study whether increased reagent concentration would also lead to higher yield. (This would need to be costed carefully since increasing reagent concentration is increasing expense.) The best support for considering such options is the interaction plot.

Exercise 2: Corrosion Study

Analyze for all effects, including curvature. State conclusions and recommendations for further experimentation.

A research group has been charged with developing a steel with better corrosion resistance than the current product. Previous experimentation has established the chemical composition of all but two elements. Chromium and nickel levels have been narrowed but not finalized. The team wishes to evaluate: chrome with levels of .04 percent and .10 percent; nickel with levels of .10 percent and .20 percent. The design was a 2^2 factorial with three replicates. Since there was a concern over non-linearity, five center points were run at the factor settings of Cr at .07 percent and Ni at .15 percent. The order of the seventeen experimental samples was randomized, and the response is the weight loss in an acid bath for a fixed time period. The objective is to minimize the weight loss. Data are shown in Table A.5.

TABLE A.5. Data for weight loss by corrosion.

			A: Chromium .04% (−)		A: Chromium .10% (+)	
B: Nickel	.10%	(−)	4.1 4.9 4.3	$\bar{Y} = 4.43$ $S^2 = .173$	5.5 4.0 4.3	$\bar{Y} = 4.60$ $S^2 = .630$
	.20%	(+)	9.6 8.7 10.1	$\bar{Y} = 9.47$ $S^2 = .503$	8.2 7.4 8.7	$\bar{Y} = 8.10$ $S^2 = .430$

Center points: 9.5, 9.0, 8.6, 8.7, 8.1 {$\bar{Y} = 8.78$; $S^2 = .267$}

1. *Calculate the effects* (see Table A.6).

TABLE A.6. Exercise 2 in spreadsheet format.

	A	B	AB	$\bar{Y}$	S^2
1	−	−	+	4.43	.173
2	+	−	−	4.60	.630
3	−	+	−	9.47	.503
4	+	+	+	8.10	.430
ΣY_+	12.7	17.57	12.53		
ΣY_-	13.9	9.03	14.07		
$\bar{Y}_+$	6.35	8.79	6.27		
$\bar{Y}_-$	6.95	4.52	7.04		
Effect	−.60	4.27	−.77		

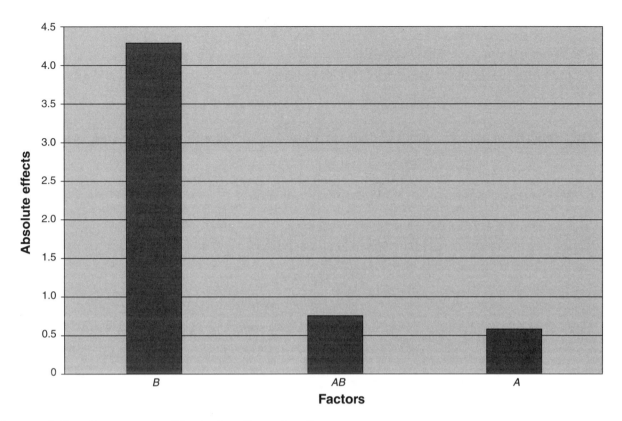

FIGURE A.5. Pareto of effects for Exercise 2.

2. *Calculate the Pareto of effects* (see Figure A.5).
3. *Experimental Error.*

   Since there are center points, additional degrees of freedom and a better estimate of experimental error are possible by modifying the usual procedures. If there were no center points, the usual procedures would be applied.

   a. Calculate and average the variances of the base four runs. Then calculate the variance at the center point.

   $$S_b^2 = (.173 + .630 + .503 + .430)/4 = .434$$
   $$S_c^2 = .267$$

   b. Calculate the degrees of freedom for the base design and for the center points.

   $$df_b = (\text{\# of runs}) \times (\text{\# of replicates} - 1) = (4)(3 - 1) = 8$$
   $$df_c = (\text{\# of replicates at the center} - 1) = 5 - 1 = 4$$

   c. Calculate S_e^2 as the weighted average of the two variance estimates, weighting each by its degrees of freedom.

   $$S_e^2 = [(df_b \times S_b^2) + (df_c \times S_c^2)]/(df_b + df_c) = [(8)(.434) + (4)(.267)]/12 = 4.54/12 = .378$$
   $$S_e = \sqrt{.378} = .62$$

4. *Calculate the standard deviation of the effects.*

$$S_{eff} = .62\sqrt{4/N} = .62\sqrt{4/12} = .36$$

5. *t-statistic.*

$$\text{Total degrees of freedom} = df_b + df_c = 8 + 4 = 12$$

For $\alpha = .05$ and $df = 12$, $t = 2.18$.

6. *Decision limits* (refer to Figure A.6).

$$DL = \pm t \times S_{eff} = \pm (2.18)(.36) = \pm .79$$

This indicates that nickel should be set at its low (.10%) value to minimize weight loss.

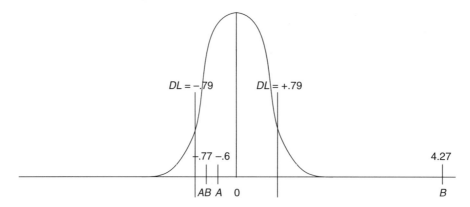

FIGURE A.6. Decision limits for effects.

7. *Plot significant effects* (see Figure A.7).

These results indicate that nickel should be set at the low (.10 percent) level to achieve minimum weight loss.

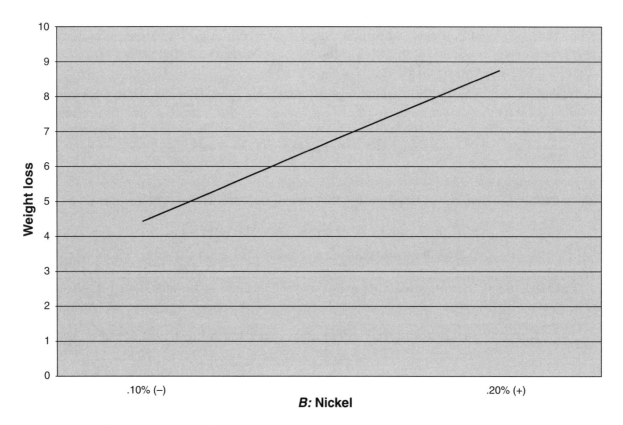

FIGURE A.7. Effect of nickel (B).

8. *Model.*

    The model is

    $$\hat{Y} = \overline{\overline{Y}} + (E(B)/2)B = 6.65 + (4.27/2)B = 6.65 + 2.14B$$

    However, before continuing, the nonlinearity check must be made.

    a. *Define the effect of nonlinearity.*

    $$E(\text{nonlinearity}) = \overline{\overline{Y}} - \overline{Y}_{center} = 6.65 - 8.78 = -2.13$$

    b. *Define the standard deviation to test nonlinearity.*

    $$S_{nonlin} = S_e \sqrt{1/N + 1/C}$$

    where N it the total number of trials not at the center and where C is the number of trials at the center.

    $$S_{nonlin} = .62 \sqrt{1/12 + 1/5} = .33$$

    c. *Calculate the decision limits for nonlinearity* (see Figure A.8).

    $$DL = \pm t \times S_{nonlin} = \pm (2.18)(.33) = \pm .72$$

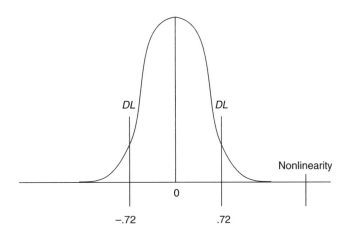

FIGURE A.8. Decision limits for nonlinearity.

Nonlinearity is significant. This means that the linear model cannot be used to optimize the process and that further defining of this process will require experimentation at three levels or more. It is safe to state only that the .10 percent nickel level is better than the .20 percent level, but we do not know what happens in between; we cannot interpolate. Since we do not know where the curvature is, we also cannot be sure of the conclusion that factor A (chromium) has no impact. We know only that these two specific levels are not significantly different in their responses.

Exercise 3: Ink Transfer

Analyze the following experiment for effects and for variation. State conclusions and recommendations for best operating conditions.

In an experiment to evaluate ink transfer when printing on industrial wrapping paper, the following three factors were used in a factorial design:

Factor	Low Level (−)	High Level (+)
A: Roll Type	Light	Heavy
B: Paper Route	Over Idler Roll	Direct
C: Drying	Fan Only	Fan Plus Heater

The ink transfer was measured by averaging inspector ratings during the processing of a roll of paper. The objective was to minimize this rating. Three independent rollings were made for each run. The order of the 24 trials was randomized. The analysis table is shown in Table A.7.

TABLE A.7. Analysis table for ink transfer exercise.

Run	A	B	C	AB	AC	BC	ABC	$\bar{Y}$	S^2
1	−	−	−	+	+	+	−	10.00	1.00
2	+	−	−	−	−	+	+	18.33	4.35
3	−	+	−	−	+	−	+	11.50	1.50
4	+	+	−	+	−	−	−	20.00	1.80
5	−	−	+	+	−	−	+	5.00	12.50
6	+	−	+	−	+	−	−	21.67	11.75
7	−	+	+	−	−	+	−	13.33	9.50
8	+	+	+	+	+	+	+	23.33	7.87
ΣY_+	83.33	68.16	63.33	58.33	66.50	64.99	58.16		
ΣY_-	39.83	55.00	59.83	64.83	56.66	58.17	65.00		
$\bar{Y}_+$	20.83	17.04	15.83	14.58	16.63	16.25	14.54		
$\bar{Y}_-$	9.96	13.75	14.96	16.21	14.17	14.54	16.25		
Effect	10.87	3.29	.87	−1.63	2.46	1.71	−1.71		
$\bar{S}_+^2$	6.44	5.17	10.41	5.79	5.53	5.68	6.56		
$\bar{S}_-^2$	6.13	7.40	2.16	6.78	7.04	6.89	6.01		
F	1.05	1.43	4.82	1.17	1.27	1.21	1.09		

1. All effects are calculated in the analysis table.
2. Create the Pareto chart of effects (see Figure A.9).

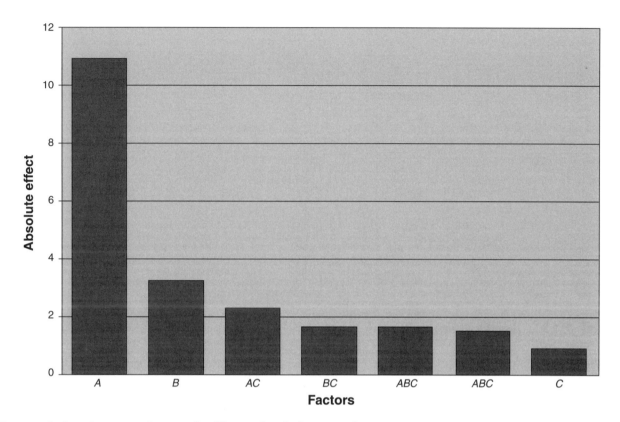

FIGURE A.9. Pareto chart of effects for ink transfer.

3. Calculate S_e.

$$S_e^2 = \{1.00 + 4.35 + ... + 7.87\}/8 = 6.28$$
$$S_e = \sqrt{6.28} = 2.51$$

4. Calculate S_{eff}.

$$S_{\mathit{eff}} = S_e \sqrt{4/N} = 2.51\sqrt{(4/24)} = 1.03$$

5. Determine degrees of freedom and t-statistic.

df = (number of runs) (number of replicates per run – 1) = (8) (3 – 1) = 16

For α = .05 and df = 16, t = 2.12.

6. Compute decision limits (see Figure A.10).

$$DL = \pm t \times S_{eff} = \pm (2.12)(1.03) = \pm 2.18$$

These results indicate that A (roll type), B (paper route), and AC (roll type-drying) interaction are significant.

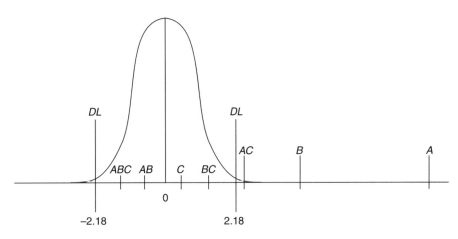

FIGURE A.10. Decision limits for ink transfer.

7. Plot the significant effects.

Plots are needed for AC interaction and B. (Since A is involved with the interaction, a main effect plot for A is not required.) The data must be condensed into a 2 × 2 table in A and C to plot the interaction (see Table A.8). The interpretation of an interaction is prone to error without a plot to provide guidance (see Figures A.11 and A.12).

TABLE A.8. Data for Exercise 3.

		A (−)	A (+)
C	(−)	10.00 11.50 Y = 10.75	18.33 20.00 Y = 19.17
	(+)	5.00 13.33 Y = 9.17	21.67 23.33 Y = 22.50

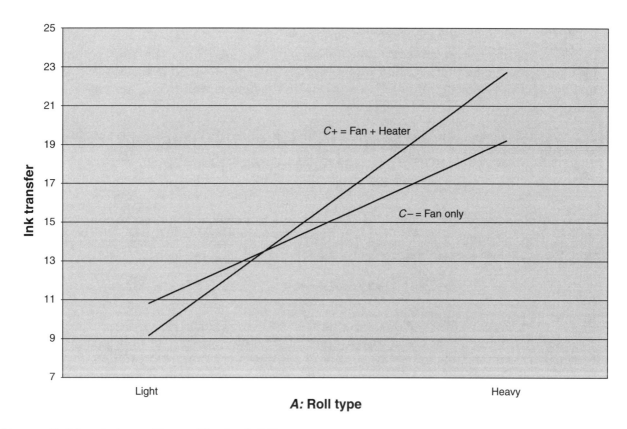

FIGURE A.11. Interaction effect of *AC*.

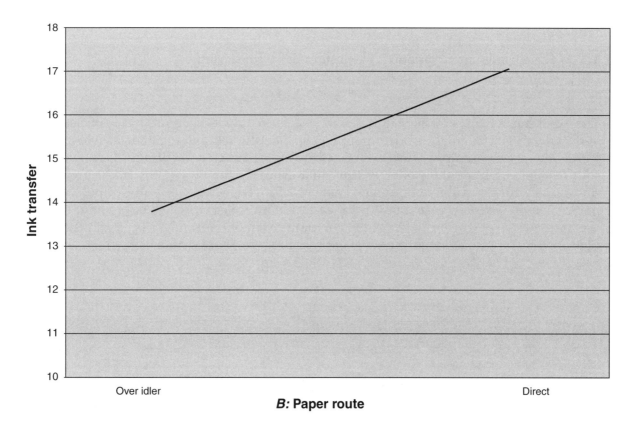

FIGURE A.12. Effect of *B*.

8. Create the model and the conclusions.

   In generating the model, remember the *hierarchy rule* for modeling. This rule requires that the individual main effects of a significant interaction must be included in the model even if those main effects were not significant. Since the AC interaction is significant, then main effects for both A and C must be in the model even though C was not significant.

   $$\hat{Y} = 15.40 + [E(A)/2]A + [E(B)/2]B + [E(C)/2]C + [E(AC)/2]AC$$
   $$= 15.40 + [10.87/2]A + [3.29/2]B + [.87/2]C + [2.46/2]AC$$
   $$= 15.40 + 5.44A + 1.65B + .44C + 1.23AC$$

   Since the objective is to minimize the response (ink transfer), the best condition would be A– (light roll), C+ (drying with fan plus heater) and B– (routing over the idler roll). At that condition, the model projects an ink transfer value of

   $$\hat{Y} = 15.40 + 5.44(-1) + 1.65(-1) + .44\ (+1) + 1.23\ (-1)(+1) = 7.52$$

9. Review the variation analysis.

   To assess the variation analysis, the degrees of freedom are split in half to determine the F-criterion from the table for α = 10 percent (5 percent in the upper tail but let's not slide into that precipice here!). $F_{(8,8)}$ = 3.44. Review of the F row in the variation part of the table indicates that only factor C (drying) has significant influence on the variation. By checking the S_+^2 and S_-^2 for factor C, we find that S_-^2 is the smaller. Thus, C– or drying with the fan only will provide much greater consistency in ink transfer.

   This is a conflict: we need to dry with fan and heater to achieve minimum ink transfer.

   $$S_e = \sqrt{10.41} = 3.23$$

   Yet, to achieve minimum variation, we need to dry with fan only.

   $$S_e = \sqrt{2.16} = 1.47$$

   The experimenters must resolve this conflict to achieve the best compromise between average and variability. This might lead to defining further studies that expand on what has been learned here. (For example, why does use of the heater increase variability? Is an even lighter roll feasible?) By using the model, we find that using the fan only (C = –1), the projected ink transfer is 9.10 versus the optimum estimate of 7.52. As a rough indicator of process performance, the projected average plus 3 S_e for each case indicates a process upper limit of 17.2 for the minimum average and 13.51 for the alternative (fan only). It might be worth the sacrifice of a small increase in average to reduce the variability by more than half.

Exercise 4: Nail Pull Study

A team is trying to improve the strength characteristics of a wall board as measured by the force required to pull a nail through the material. The board is made up of layers of paper pasted together and pressed. The final list of factors is listed following showing the low versus high levels. The objective is to maximize the response. There were two replicates of each run, each being a completely random and independent setup.

- A: Paste temperature—120° F vs. 150° F
- B: Roll pressure—40 percent vs. 80 percent
- C: Amount of sizing—.5 percent vs. 1.0 percent
- D: Paste types—X vs. Y
- E: Paper moisture—4 percent vs. 8 percent
- F: Cure time—5 days vs. 10 days
- G: Mill speed—200 fpm vs. 250 fpm

The factors vary in their expense to the process:

1. Extending cure time creates a massive and expensive storage problem.
2. Mill speed is directly correlated with productivity.
3. High-moisture paper is cheaper than low-moisture paper.
4. Increasing the amount of sizing adds to the cost of board.
5. Temperature and paste type are very minor in their cost impact.
6. Roll pressure has no cost impact.

Analyze and make recommendations to management. The analysis table is shown in Table A.9 on page 82.

1. The effects are shown in the analysis table.
2. Pareto chart of effects (see Figure A.13 on page 82).

   Based on the Pareto chart, it is apparent that factor C (amount of sizing) is a significant effect. A decision on the next group of factors will require the formal significance test.

3. Standard deviation of the effects.

$$S_e = \sqrt{(\Sigma\ S_i^2 / k)} = \sqrt{(.61 + 38.69 + 3.65 + \ldots\ldots + 12.50)/8} = \sqrt{22.26} = 4.72$$

4. Standard deviation of the effects.

$$S_{eff} = S_e \sqrt{4/N} = 4.72\ (\sqrt{4/16}) = 2.36$$

5. Degrees of freedom and t-statistic.

$$df = (\text{number of replicates} - 1) \times (\text{number of runs}) = (2 - 1) \times (8) = 8$$

For $\alpha = .05$ and $df = 8$, from Table A.1, the table of t-values, t = 2.31.

TABLE A.9. Analysis table for exercise 4.

	A	B	C	D	E	F	G	$\bar{Y}$	S^2
1	+	−	−	+	−	+	+	59.75	.61
2	+	+	−	−	+	−	+	56.20	38.69
3	+	+	+	−	−	+	−	70.15	3.65
4	−	+	+	+	−	−	+	72.60	.02
5	+	−	+	+	+	−	−	70.90	7.24
6	−	+	−	+	+	+	−	55.55	73.27
7	−	−	+	−	+	+	+	63.20	42.10
8	−	−	−	−	−	−	−	50.00	12.50
$\sum Y_+$	257.00	254.50	276.85	258.80	245.85	248.65	251.75		
$\sum Y_-$	241.35	243.85	221.50	234.55	252.50	249.70	246.60		
$\bar{Y}_+$	64.25	63.63	69.21	64.70	61.46	62.16	62.94		
$\bar{Y}_-$	60.34	60.96	55.37	59.89	63.13	62.43	61.65		
Effect	3.91	2.67	13.84	4.81	−1.67	−0.27	1.29		
$\bar{S}_+^2$	12.55	28.91	13.25	20.29	40.33	29.91	20.36		
$\bar{S}_-^2$	31.97	15.61	31.27	24.24	4.20	14.61	24.17		
F	2.55	1.85	2.36	1.19	9.60	2.05	1.19		
Aliases	−BD	−AD	−AG	−AB	−AF	−AE	−AC		
	−CG	−CE	−BE	−CF	−BC	−BG	−BF		
	−EF	−FG	−DF	−EG	−DG	−CD	−DE		

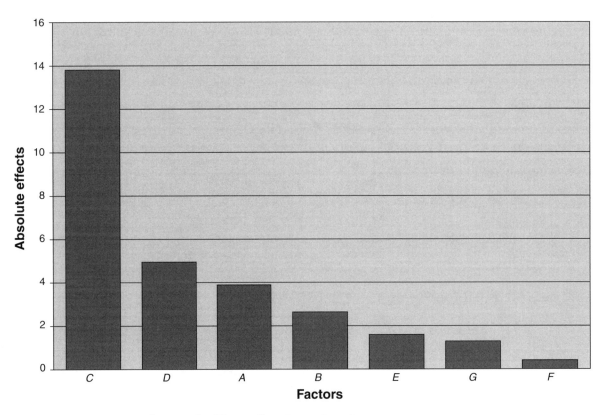

FIGURE A.13. Pareto chart of effects for Exercise 4.

6. Decision limits (see Figure A.14).

$$DL = \pm t \times S_{eff} = \pm (2.31 (2.36)) = \pm 5.45$$

Based on the decision limits, only factor C (amount of sizing) is significant.

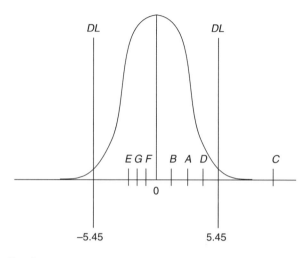

FIGURE A.14. Decision limits.

7. Plot of significant effect (see Figure A.15).

Since the objective is to maximize the response (force to pull a nail through the board), the graph indicates that the best condition is at the high level of factor C, amount of sizing at one percent.

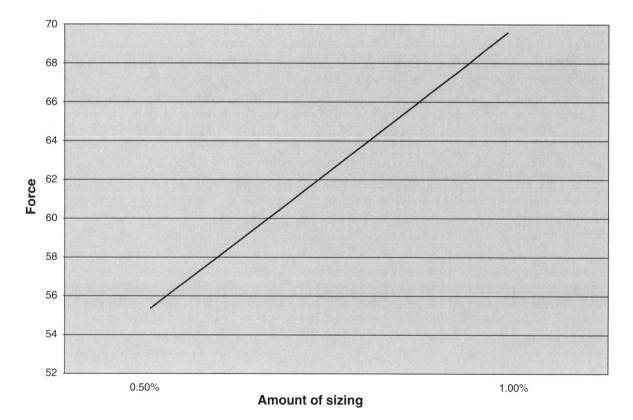

FIGURE A.15. Effect of amount of sizing.

8. Model and conclusions.

$$\hat{\bar{Y}} = \bar{\bar{Y}} + \{\frac{E(C)}{2}\}C = 62.29 + (13.84/2)C = 62.29 + 6.92C$$

If we substitute $C = +1$ to represent the high level of sizing, the projection is 69.21.

Next, the variation analysis is reviewed. The degrees of freedom are divided in half to determine the F-criterion from Table A.2, the F-table. $F_{(4,4)} = 6.39$. Comparing the results of the F-tests in the variation analysis to this reference value determined that factor E, paper moisture, has significant impact on the inherent or experimental variation since its test value exceeded the reference value. Checking the individual estimates of the variances, the low level of factor E provides the smaller variance, 4.20. The estimate of the standard deviation of the experiment (S_e) for that condition is $\sqrt{4.20} = 2.05$.

The conclusions are

a. The response is maximized by using the larger amount of sizing. This is more expensive, and the cost increase must be evaluated against the gain in response.

b. Lower moisture paper will greatly reduce the variation in the wall board and is again more expensive.

c. The other factors can be set at the most economic conditions. Specifically, the shorter cure time and higher mill speeds should be used since they provide major cost reductions.

d. Further factors should be sought that could be used to offset the cost of higher sizing usage and lower moisture paper.

Note the aliases of both C and E:

$$C = -AG = -BE = -DF$$

$$E = -AF = -BC = -DG$$

The technical experts (be careful here!) see only *DF* (paste type and cure time) and *AF* (paste temperature and cure time) as potential interactions. These two significant factors should be studied in a small verification run to confirm their results. An alternative would be to perform a reflection of this design to assure that interactions are not involved.

Exercise 5: The Lawn Fanatics

A lawn service wanted to analyze the factors that lead to a high-growth lawn. They had a very homogeneous five-acre plot that was used for the experiment. They determined the following ten factors to investigate:

A: Amount of water per week—1" vs. 2"

B: Amount of lime—2# per plot vs. 5#

C: Amount of nitrogen—.10 percent vs. .30 percent

D: Amount of phosphorous—.1 percent vs. .2 percent

E: Grass type—Brand *X* vs. Brand *Y*

F: Blade condition—dull vs. sharp

G: Grass height after mowing—2 inches vs. 2.5 inches

H: Insecticide—none vs. regular application

I: Weed treatment—none vs. regular application

J: Time of day of watering—5 A.M. vs. 4 P.M.

The design was a 12-run Plackett-Burman with reflection. The field was divided into 24 plots, each two-tenths of an acre. The 24 trials were assigned to the plots by drawing numbers from a box. The response was the average weight of the grass clippings from the plots, averaged over six mowings. The mowings were seven days apart. The objective is to maximize the response. The analysis table is shown in Table A.10. Analyze, generate conclusions and recommendations, and critique the experiment.

1. The effects are calculated in the analysis table (Table A.10 on page 86).

2. Pareto chart of effects (see Figure A.16).

   Since there is no replication, the Pareto chart must be used to draw conclusions. From that chart, it is apparent that factors *A* (water), *B* (lime), and *C* (nitrogen) are significant. All other factors are minor in comparison to these three.

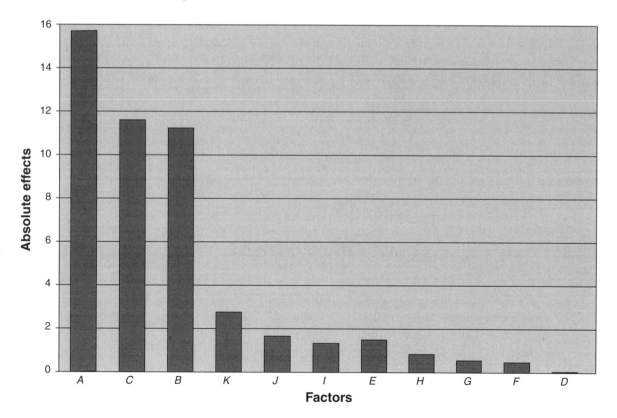

FIGURE A.16. Pareto chart of effects for Example 5.

TABLE A.10. Analysis table for Exercise 5.

	A	B	C	D	E	F	G	H	I	J	K	Y
1	+	−	+	−	−	−	+	+	+	−	+	53.0
2	+	+	−	+	−	−	−	+	+	+	−	63.2
3	−	+	+	−	+	−	−	−	+	+	+	48.2
4	+	−	+	+	−	+	−	−	−	+	+	58.7
5	+	+	−	+	+	−	+	−	−	−	+	53.4
6	+	+	+	−	+	+	−	+	−	−	−	71.6
7	−	+	+	+	−	+	+	−	+	−	−	50.5
8	−	−	+	+	+	−	+	+	−	+	−	44.3
9	−	−	−	+	+	+	−	+	+	−	+	31.2
10	+	−	−	−	+	+	+	−	+	+	−	48.4
11	−	+	−	−	−	+	+	+	−	+	+	40.7
12	−	−	−	−	−	−	−	−	−	−	−	25.0
13	−	+	−	+	+	+	−	−	−	+	−	45.4
14	−	−	+	−	+	+	+	−	−	−	+	46.4
15	+	−	−	+	−	+	+	+	−	−	−	44.2
16	−	+	−	−	+	−	+	+	+	−	−	47.1
17	−	−	+	−	+	−	+	+	+	+	−	43.2
18	−	−	−	+	−	−	+	−	+	+	+	30.9
19	+	−	−	−	+	−	−	+	−	+	+	48.0
20	+	+	−	−	−	+	−	−	+	−	+	54.8
21	+	+	+	−	−	−	+	−	−	+	−	76.0
22	−	+	+	+	−	−	−	+	−	−	+	55.7
23	+	−	+	+	+	−	−	−	+	−	−	60.3
24	+	+	+	+	+	+	+	+	+	+	+	65.0
ΣY_+	696.6	671.6	672.9	602.8	609.3	600.1	599.9	607.2	595.8	612.0	586.0	
ΣY_-	508.6	533.6	532.3	602.4	595.9	605.1	605.3	598.0	609.4	593.2	619.2	
Y_+	58.05	55.97	56.08	50.23	50.78	50.01	49.99	50.60	49.65	51.00	48.83	
Y_-	42.38	44.47	44.36	50.20	49.66	50.43	50.44	49.83	50.78	49.43	51.60	
Effect	15.67	11.50	11.72	.03	1.12	−.42	−.45	.77	−1.13	1.57	−2.77	

3. Standard deviation of the experiment: not available without replication.

4. Standard deviation of the effects: not available.

5. Degrees of freedom and t-statistic: not available.

6. Decision limits: not available.

7. Plot significant effects (see Figures A.17, A.18, and A.19):

8. Model and conclusions.

   The results, based on the Pareto chart, indicate that A (water), C (nitrogen), and B (lime) are significant factors that can control lawn growth. Did you also notice that there were only ten factors? Since K was not needed, it was a *dummy* factor, representing the inherent variation of the experiment. (Had the experiment not been reflected, it would also have represented the partial interactions that would be confounded with that factor.) These results appear very straightforward; they will not always be so clear in a nongeometric design. A prudent decision

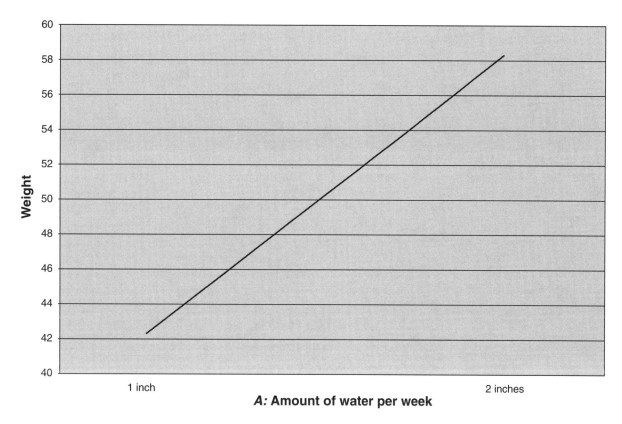

FIGURE A.17. Effect of A (water).

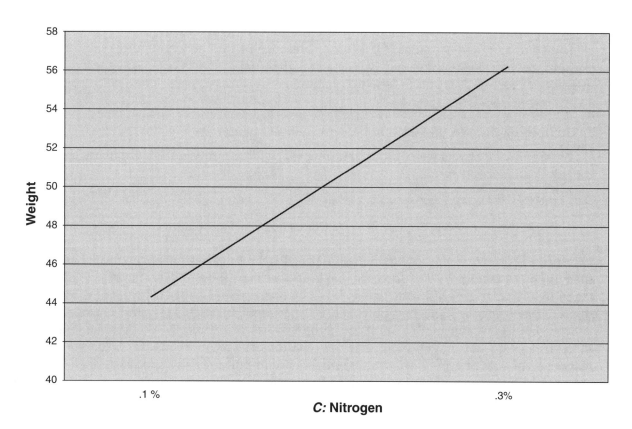

FIGURE A.18. Effect of C (nitrogen).

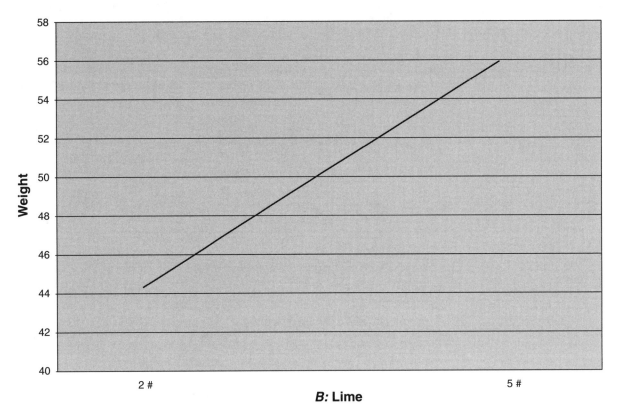

FIGURE A.19. Effect of *B* (lime).

was made to reflect this experiment rather than replicate. This assures that the main effects are clear of two-factor interactions. Otherwise, presence of interactions could make the conclusions more risky. Since the heredity rule states that major effects often have interactions, the interactions for *AB, AC,* and *BC* should be checked by

a. Calculating an analysis column for each
b. Computing the effect for each in the spreadsheet fashion

No significant effects were found by doing so; i.e., those effects were about the same size as the other *noise* effects.

The model is

$$Y = \bar{Y} + [E(A)/2]A + [E(B)]B + [E(C)]C = 50.22 + 7.84\,A + 5.75\,B + 5.86\,C$$

Since the objective was to maximize the growth, the three factors in the model are all set to their high (+) condition. The forecast maximum is 69.67. These three factors would now be taken to a smaller experiment to assess linearity and optimality. Their levels in the next study might be set using the high conditions of this experiment as the center point. The nonsignificant factors should be reviewed to see if any economic decisions can be made since the levels of those factors do not matter. Note that the decisions we are making are based solely on the response variable of growth. If other responses are important, they should be included in the study. Adding responses is like having free experiments; the benefits are available with only their measurement required.

The major critique of this experiment is the use of reflection instead of replication. Lack of replication was not a problem in making decisions about significance. However, the complex confounding pattern of a nongeometric design could have clouded the conclusions had the reflection not been run to assure that main effects were not confounded with partial interactions. It is always worth checking the significant factors for interactions in such designs, but the conclusions about them should always be verified in the succeeding experiment.

Some criticism might be addressed at the experimental procedure. Some factors were immediate in their implementation, such as grass height after mowing, blade condition, etc. Others could have a time lag, e.g., the chemicals and water application. While the data were averaged over six weeks, an initial delay might be beneficial to provide time for these slower-reacting factors to have an impact on growth. The data averaged over four weeks, allowing the initial two weeks for the chemicals to generate any reaction, might have been more appropriate.

TABLE A.11. Analysis table for 2^3 factorial.

	A	B	C	AB	AC	BC	$\bar{Y}$	S^2
1	−	−	−	+	+	+		
2	+	−	−	−	−	+		
3	−	+	−	−	+	−		
4	+	+	−	+	−	−		
5	−	−	+	+	−	−		
6	+	−	+	−	+	−		
7	−	+	+	−	−	+		
8	+	+	+	+	+	+		
$\sum Y_+$								
$\sum Y_-$								
$\bar{Y}_+$								
$\bar{Y}_-$								
Effect								

TABLE A.12. Analysis table for 2^3 factorial.

	A	B	C	AB	AC	BC	$\overline{Y}$	S^2
1	−	−	−	+	+	+		
2	+	−	−	−	−	+		
3	−	+	−	−	+	−		
4	+	+	−	+	−	−		
5	−	−	+	+	−	−		
6	+	−	+	−	+	−		
7	−	+	+	−	−	+		
8	+	+	+	+	+	+		
ΣY_+								
ΣY_-								
$\overline{Y}_+$								
$\overline{Y}_-$								
Effect								
$\overline{S}_+^2$								
$\overline{S}_-^2$								
F								

TABLE A.13. Design matrix for ½ fractional factorial for four factors.

	A	B	C	D
1	−	−	−	−
2	+	−	−	+
3	−	+	−	+
4	+	+	−	+
5	−	−	+	+
6	+	−	+	−
7	−	+	+	−
8	+	+	+	+

Confounding pattern: $AB = CD$
$AC = BD$
$AD = BC$

TABLE A.14. Analysis table for $\frac{1}{2}$ fractional factorial with four factors.

	A	B	C	D	AB=CD	AC=BD	AD=BC	$\overline{Y}$	S^2
1	−	−	−	−	+	+	+		
2	+	−	−	+	−	−	+		
3	−	+	−	+	−	+	−		
4	+	+	−	−	+	−	−		
5	−	−	+	+	+	−	−		
6	+	−	+	−	−	+	−		
7	−	+	+	−	−	−	+		
8	+	+	+	+	+	+	+		
$\sum Y_+$									
$\sum Y_-$									
$\overline{Y}_+$									
$\overline{Y}_-$									
Effect									
$\overline{S}_+^2$									
$\overline{S}_-^2$									
F									

TABLE A.15. Design matrix for $\frac{1}{4}$ fractional factorial for five factors.

	A	B	C	D	E
1	−	−	−	+	+
2	+	−	−	−	−
3	−	+	−	−	+
4	+	+	−	+	−
5	−	−	+	+	−
6	+	−	+	−	+
7	−	+	+	−	−
8	+	+	+	+	+
Confounding:	BD CE	AD	AE	AB	AC

TABLE A.16. Analysis table for $\frac{1}{4}$ fractional factorial with five factors.

	A	B	C	D	E	BC=DE	BE=CD	$\overline{Y}$	S^2
1	−	−	−	+	+	+	−		
2	+	−	−	−	−	+	+		
3	−	+	−	−	+	−	+		
4	+	+	−	+	−	−	−		
5	−	−	+	+	−	−	+		
6	+	−	+	−	+	−	−		
7	−	+	+	−	−	+	−		
8	+	+	+	+	+	+	+		
$\sum Y_+$									
$\sum Y_-$									
$\overline{Y}_+$									
$\overline{Y}_-$									
Effect									
$\overline{S}_+^2$									
$\overline{S}_-^2$									
F									

TABLE A.17. Design table for $\frac{1}{2}$ fractional factorial with five factors.

	A	B	C	D	E
1	−	−	−	−	+
2	+	−	−	−	−
3	−	+	−	−	−
4	+	+	−	−	+
5	−	−	+	−	−
6	+	−	+	−	+
7	−	+	+	−	+
8	+	+	+	−	−
9	−	−	−	+	−
10	+	−	−	+	+
11	−	+	−	+	+
12	+	+	−	+	−
13	−	−	+	+	+
14	+	−	+	+	−
15	−	+	+	+	−
16	+	+	+	+	+

TABLE A.18. Analysis table for $\frac{1}{2}$ fractional factorial with five factors.

	A	B	C	D	E	AB	AC	AD	AE	BC	BD	BE	CD	CE	DE	$\bar{Y}$	S^2
1	−	−	−	−	+	+	+	+	+	+	+	−	+	−	−		
2	+	−	−	−	−	−	−	−	−	+	+	+	+	+	+		
3	−	+	−	−	−	−	+	+	+	−	−	−	+	+	+		
4	+	+	−	−	+	+	−	−	−	−	−	+	+	−	−		
5	−	−	+	−	−	+	−	+	+	−	+	+	−	−	+		
6	+	−	+	−	+	−	+	−	−	−	+	−	−	+	−		
7	−	+	+	−	+	−	−	+	+	+	−	+	−	+	−		
8	+	+	+	−	−	+	+	−	−	+	−	−	−	−	+		
9	−	−	−	+	−	+	+	−	+	+	−	+	−	+	−		
10	+	−	−	+	+	−	−	+	−	+	−	−	−	−	+		
11	−	+	−	+	+	−	+	−	+	−	+	+	−	−	+		
12	+	+	−	+	−	+	−	+	−	−	+	−	−	+	−		
13	−	−	+	+	+	+	−	−	+	−	−	−	+	+	+		
14	+	−	+	+	−	−	+	+	−	−	−	+	+	−	−		
15	−	+	+	+	−	−	−	−	+	+	+	−	+	−	−		
16	+	+	+	+	+	+	+	+	−	+	+	+	+	+	+		
ΣY_+																	
ΣY_-																	
$\bar{Y}_+$																	
$\bar{Y}_-$																	
Effect																	
$\bar{S}_+^2$																	
$\bar{S}_-^2$																	
F																	

TABLE A.19. Confounding pattern for eight-run Plackett-Burman design.

Factors	A	B	C	D	E	F	G
Two-Factor	–BD	–AD	–AG	–AB	–AF	–AE	–AC
Interactions	–CG	–CE	–BE	–CF	–BC	–BG	–BF
	–EF	–FG	–DF	–EG	–DG	–CD	–DE

TABLE A.20. Confounding pattern for 16-run Plackett-Burman design.

A	B	C	D	E	F	G	H	I	J	K	L	M	N	O
–BE	–AE	–AI	–AO	–AB	–AK	–AN	–AJ	–AC	–AH	–AF	–AM	–AL	–AG	–AD
–CI	–CF	–BF	–BJ	–CK	–BC	–BL	–BO	–BK	–BD	–BI	–BG	–BN	–BM	–BH
–DO	–DJ	–DG	–CG	–DH	–DL	–CD	–CM	–DN	–CL	–CE	–CJ	–CH	–CO	–CN
–FK	–GL	–EK	–EH	–FI	–EI	–EM	–DE	–EF	–EO	–DM	–DF	–DK	–DI	–EJ
–GN	–HO	–HM	–FL	–GM	–GJ	–FJ	–FN	–GO	–FG	–GH	–EN	–EG	–EL	–FM
–HJ	–IK	–JL	–IN	–JO	–HN	–HK	–GK	–HL	–IM	–JN	–HI	–FO	–FH	–GI
–LM	–MN	–NO	–KM	–LN	–MO	–IO	–IL	–JM	–KN	–LO	–KO	–IJ	–JK	–KL

TABLE A.21. Analysis table for eight-run Plackett-Burman design.

	A	B	C	D	E	F	G	$\bar{Y}$	S^2
1	+	−	−	+	−	+	+		
2	+	+	−	−	+	−	+		
3	+	+	+	−	−	+	−		
4	−	+	+	+	−	−	+		
5	+	−	+	+	+	−	−		
6	−	+	−	+	+	+	−		
7	−	−	+	−	+	+	+		
8	−	−	−	−	−	−	−		
ΣY_+									
ΣY_-									
$\bar{Y}_+$									
$\bar{Y}_-$									
Effect									
$\bar{S}_+^2$									
$\bar{S}_-^2$									
F									

TABLE A.22. Analysis table for reflection of eight-run Plackett-Burman design.

	A	B	C	D	E	F	G	$\bar{Y}$	S^2
1	−	+	+	−	+	−	−		
2	−	−	+	+	−	+	−		
3	−	−	−	+	+	−	+		
4	+	−	−	−	+	+	−		
5	−	+	−	−	−	+	+		
6	+	−	+	−	−	−	+		
7	+	+	−	+	−	−	−		
8	+	+	+	+	+	+	+		
ΣY_+									
ΣY_-									
$\bar{Y}_+$									
$\bar{Y}_-$									
Effect									
$\bar{S}_+^2$									
$\bar{S}_-^2$									
F									

TABLE A.23. Analysis table for eight-run Plackett-Burman design with reflection.

	A	B	C	D	E	F	G	$\overline{Y}$	S^2
1	+	−	−	+	−	+	+		
2	+	+	−	−	+	−	+		
3	+	+	+	−	−	+	−		
4	−	+	+	+	−	−	+		
5	+	−	+	+	+	−	−		
6	−	+	−	+	+	+	−		
7	−	−	+	−	+	+	+		
8	−	−	−	−	−	−	−		
9	−	+	+	−	+	−	−		
10	−	−	+	+	−	+	−		
11	−	−	−	+	+	−	+		
12	+	−	−	−	+	+	−		
13	−	+	−	−	−	+	+		
14	+	−	+	−	−	−	+		
15	+	+	−	+	−	−	−		
16	+	+	+	+	+	+	+		
ΣY_+									
ΣY_-									
$\overline{Y}_+$									
$\overline{Y}_-$									
Effect									
$\overline{S}_+^2$									
$\overline{S}_-^2$									
F									

TABLE A.24. Analysis table for 12-run Plackett-Burman design.

	A	B	C	D	E	F	G	H	I	J	K	$\bar{Y}$	S^2
1	+	−	+	−	−	−	+	+	+	−	+		
2	+	+	−	+	−	−	−	+	+	+	−		
3	−	+	+	−	+	−	−	−	+	+	+		
4	+	−	+	+	−	+	−	−	−	+	+		
5	+	+	−	+	+	−	+	−	−	−	+		
6	+	+	+	−	+	+	−	+	−	−	−		
7	−	+	+	+	−	+	+	−	+	−	−		
8	−	−	+	+	+	−	+	+	−	+	−		
9	−	−	−	+	+	+	−	+	+	−	+		
10	+	−	−	−	+	+	+	−	+	+	−		
11	−	+	−	−	−	+	+	+	−	+	+		
12	−	−	−	−	−	−	−	−	−	−	−		
ΣY_+													
ΣY_-													
$\bar{Y}_+$													
$\bar{Y}_-$													
Effect													
$\bar{S}_+^2$													
$\bar{S}_-^2$													
F													

TABLE A.25. Analysis table for 12-run Plackett-Burman design (unreplicated).

	A	B	C	D	E	F	G	H	I	J	K	Y
1		−	+	−	−	−	+	+	+	−	+	
2	+	+	−	+	−	−	−	+	+	+	−	
3	−	+	+	−	+	−	−	−	+	+	+	
4	+	−	+	+	−	+	−	−	−	+	+	
5	+	+	−	+	+	−	+	−	−	−	+	
6	+	+	+	−	+	+	−	+	−	−	−	
7	−	+	+	+	−	+	+	−	+	−	−	
8	−	−	+	+	+	−	+	+	−	+	−	
9	−	−	−	+	+	+	−	+	+	−	+	
10	+	−	−	−	+	+	+	−	+	+	−	
11	−	+	−	−	−	+	+	+	−	+	+	
12	−	−	−	−	−	−	−	−	−	−	−	
ΣY_+												
ΣY_-												
$\overline{Y}_+$												
$\overline{Y}_-$												
Effect												

TABLE A.26. Analysis table for 12-run Plackett-Burman design with reflection.

	A	B	C	D	E	F	G	H	I	J	K	Y	S^2
1	+	−	+	−	−	−	+	+	+	−	+		
2	+	+	−	+	−	−	−	+	+	+	−		
3	−	+	+	−	+	−	−	−	+	+	+		
4	+	−	+	+	−	+	−	−	−	+	+		
5	+	+	−	+	+	−	+	−	−	−	+		
6	+	+	+	−	+	+	−	+	−	−	−		
7	−	+	+	+	−	+	+	−	+	−	−		
8	−	−	+	+	+	−	+	+	−	+	−		
9	−	−	−	+	+	+	−	+	+	−	+		
10	+	−	−	−	+	+	+	−	+	+	−		
11	−	+	−	−	−	+	+	+	−	+	+		
12	−	−	−	−	−	−	−	−	−	−	−		
13	−	+	−	+	+	+	−	−	−	+	−		
14	−	−	+	−	+	+	+	−	−	−	+		
15	+	−	−	+	−	+	+	+	−	−	−		
16	−	+	−	−	+	−	+	+	+	−	−		
17	−	−	+	−	−	+	−	+	+	+	−		
18	−	−	−	+	−	−	+	−	+	+	+		
19	+	−	−	−	+	−	−	+	−	+	+		
20	+	+	−	−	−	+	−	−	+	−	+		
21	+	+	+	−	−	−	+	−	−	+	−		
22	−	+	+	+	−	−	−	+	−	−	+		
23	+	−	+	+	+	−	−	−	+	−	−		
24	+	+	+	+	+	+	+	+	+	+	+		
ΣY_+													
ΣY_-													
$\overline{Y}_+$													
$\overline{Y}_-$													
Effect													
$\overline{S}_+^2$													
$\overline{S}_-^2$													
F													

TABLE A.27. Analysis table for 16-run Plackett-Burman design.

	A	B	C	D	E	F	G	H	I	J	K	L	M	N	O	$\bar{Y}$	S^2
1	+	−	−	−	+	−	−	+	+	−	+	−	+	+	+		
2	+	+	−	−	−	+	−	−	+	+	−	+	−	+	+		
3	+	+	+	−	−	−	+	−	−	+	+	−	+	−	+		
4	+	+	+	+	−	−	−	+	−	−	+	+	−	+	−		
5	−	+	+	+	+	−	−	−	+	−	−	+	+	−	+		
6	+	−	+	+	+	+	−	−	−	+	−	−	+	+	−		
7	−	+	−	+	+	+	+	−	−	−	+	−	−	+	+		
8	+	−	+	−	+	+	+	+	−	−	−	+	−	−	+		
9	+	+	−	+	−	+	+	+	+	−	−	−	+	−	−		
10	−	+	+	−	+	−	+	+	+	+	−	−	−	+	−		
11	−	−	+	+	−	+	−	+	+	+	+	−	−	−	+		
12	+	−	−	+	+	−	+	−	+	+	+	+	−	−	−		
13	−	+	−	−	+	+	−	+	−	+	+	+	+	−	−		
14	−	−	+	−	−	+	+	−	+	−	+	+	+	+	−		
15	−	−	−	+	−	−	+	+	−	+	−	+	+	+	+		
16	−	−	−	−	−	−	−	−	−	−	−	−	−	−	−		
ΣY_+																	
ΣY_-																	
$\bar{Y}_+$																	
$\bar{Y}_-$																	
Effect																	
$\bar{S}_+^2$																	
$\bar{S}_-^2$																	
F																	

TABLE A.28. Analysis table for 16-run Plackett-Burman design with reflection.

	A	B	C	D	E	F	G	H	I	J	K	L	M	N	O	$\bar{Y}$	S^2
1	+	−	−	−	+	−	−	+	+	−	+	−	+	+	+		
2	+	+	−	−	−	+	−	−	+	+	−	+	−	+	+		
3	+	+	+	−	−	−	+	−	−	+	+	−	+	−	+		
4	+	+	+	+	−	−	−	+	−	−	+	+	−	+	−		
5	−	+	+	+	+	−	−	−	+	−	−	+	+	−	+		
6	+	−	+	+	+	+	−	−	−	+	−	−	+	+	−		
7	−	+	−	+	+	+	+	−	−	−	+	−	−	+	+		
8	+	−	+	−	+	+	+	+	−	−	−	+	−	−	+		
9	+	+	−	+	−	+	+	+	+	−	−	−	+	−	−		
10	−	+	+	−	+	−	+	+	+	+	−	−	−	+	−		
11	−	−	+	+	−	+	−	+	+	+	+	−	−	−	+		
12	+	−	−	+	+	−	+	−	+	+	+	+	−	−	−		
13	−	+	−	−	+	+	−	+	−	+	+	+	+	−	−		
14	−	−	+	−	−	+	+	−	+	−	+	+	+	+	−		
15	−	−	−	+	−	−	+	+	−	+	−	+	+	+	+		
16	−	−	−	−	−	−	−	−	−	−	−	−	−	−	−		
17	−	+	+	+	−	+	+	−	−	+	−	+	−	−	−		
18	−	−	+	+	+	−	+	+	−	−	+	−	+	−	−		
19	−	−	−	+	+	+	−	+	+	−	−	+	−	+	−		
20	−	−	−	−	+	+	+	−	+	+	−	−	+	−	+		
21	+	−	−	−	−	+	+	+	−	+	+	−	−	+	−		
22	−	+	−	−	−	−	+	+	+	−	+	+	−	−	+		
23	+	−	+	−	−	−	−	+	+	+	−	+	+	−	−		
24	−	+	−	+	−	−	−	−	+	+	+	−	+	+	−		
25	−	−	+	−	+	−	−	−	−	+	+	+	−	+	+		
26	+	−	−	+	−	+	−	−	−	−	+	+	+	−	+		
27	+	+	−	−	+	−	+	−	−	−	−	+	+	+	−		
28	−	+	+	−	−	+	−	+	−	−	−	−	+	+	+		
29	+	−	+	+	−	−	+	−	+	−	−	−	−	+	+		
30	+	+	−	+	+	−	−	+	−	+	−	−	−	−	+		
31	+	+	+	−	+	+	−	−	+	−	+	−	−	−	−		
32	+	+	+	+	+	+	+	+	+	+	+	+	+	+	+		
ΣY_+																	
ΣY_-																	
$\bar{Y}_+$																	
$\bar{Y}_-$																	
Effect																	
$\bar{S}_+^2$																	
$\bar{S}_-^2$																	
F																	

Table A.29. Design matrix for 20-run Plackett-Burman.

	A	B	C	D	E	F	G	H	I	J	K	L	M	N	O	P	Q	R	S
1	+	−	+	+	−	−	−	−	+	−	+	−	+	+	+	+	−	−	+
2	+	+	−	+	+	−	−	−	−	+	−	+	−	+	+	+	+	−	−
3	−	+	+	−	+	+	−	−	−	−	+	−	+	−	+	+	+	+	−
4	−	−	+	+	−	+	+	−	−	−	−	+	−	+	−	+	+	+	+
5	+	−	−	+	+	−	+	+	−	−	−	−	+	−	+	−	+	+	+
6	+	+	−	−	+	+	−	+	+	−	−	−	−	+	−	+	−	+	+
7	+	+	+	−	−	+	+	−	+	+	−	−	−	−	+	−	+	−	+
8	+	+	+	+	−	−	+	+	−	+	+	−	−	−	−	+	−	+	−
9	−	+	+	+	+	−	−	+	+	−	+	+	−	−	−	−	+	−	+
10	+	−	+	+	+	+	−	−	+	+	−	+	+	−	−	−	−	+	−
11	−	+	−	+	+	+	+	−	−	+	+	−	+	+	−	−	−	−	+
12	+	−	+	−	+	+	+	+	−	−	+	+	−	+	+	−	−	−	−
13	−	+	−	+	−	+	+	+	+	−	−	+	+	−	+	+	−	−	−
14	−	−	+	−	+	−	+	+	+	+	−	−	+	+	−	+	+	−	−
15	−	−	−	+	−	+	−	+	+	+	+	−	−	+	+	−	+	+	−
16	−	−	−	−	+	−	+	−	+	+	+	+	−	−	+	+	−	+	+
17	+	−	−	−	−	+	−	+	−	+	+	+	+	−	−	+	+	−	+
18	+	+	−	−	−	−	+	−	+	−	+	+	+	+	−	−	+	+	−
19	−	+	+	−	−	−	−	+	−	+	−	+	+	+	+	−	−	+	+
20	−	−	−	−	−	−	−	−	−	−	−	−	−	−	−	−	−	−	−

Glossary

Aliases Effects that are identical except for the sign, i.e., confounded effects.

Alpha risk The probability of erroneously claiming a difference in two averages or two variances.

Analysis matrix The table that displays the design columns and any interaction columns with the outline of the calculations to be completed.

Center points With quantitative factors, the run defined when the levels of each factor are set at the midpoint of the high and low levels.

Confounding The result of defining columns for main effects and/or interactions that are identical or exactly reversed in sign. The resulting set of confounded effects are *aliases* of one another.

Degrees of freedom A measure of the amount of information available to estimate the standard deviation of the experiment.

Design matrix The table that shows the settings for each factor for each run.

Design of experiments A family of efficient techniques for studying more than one factor in a single experiment.

Experimental error The random noise in the experiment; an estimate of the inherent variation of the process.

Factors Controllable (independent) variables that may cause a change in the response (dependent) variable.

Foldover The extension of a basic screening design by changing all signs; this assures that main effects will no longer be confounded with two-factor interactions in a combined analysis. Also called a *reflection*.

Fractional factorials Designs that contain a fraction (or power of $\frac{1}{2}$) of the number of full factorial runs. The resulting confounding is complete instead of partial. Can be effectively optimized for any number of factors.

F-test The ratio of variances used to test hypotheses on equality of variances.

Full factorial design The combination of all levels of all factors in an experiment.

Geometric screening design Screening designs that are a power of two (e.g., runs of 8, 16, 32, 64, etc.). The confounding of effects is complete. All fractional factorials are part of this family as are Plackett-Burman designs for those run sizes.

Heredity The characteristic that factors with large effects will often have a significant interaction.

Hierarchy of models The preferred practice of including all main-effect terms in a model when their interaction is significant, regardless of the significance of the main effect terms.

Interaction When the effect of factor A differs according to what the level factor of B is. A combined effect that cannot be predicted by the individual main effects.

Levels The specific settings or options for a factor in an experiment.

Main effect The average change in the response due to the change in the levels of a factor.

Nongeometric screening designs Screening designs that are a multiple of four and not a power of two (e.g., 12, 20, 24, 28, 36, etc.). The confounding of effects is partial. These designs are uniquely Plackett-Burman.

Outliers Abnormal responses resulting from special causes or uncontrolled influences that occur during an experiment.

Pareto chart A graphical tool for prioritizing effects.

Plackett-Burman designs A family of screening designs that can address $n-1$ factors with n trials so long as n is divisible by four. Designs can be geometric (the number of runs is a power of two) or nongeometric.

Refining designs Designs that optimize a small number of factors, generally using more than two levels.

Reflection The extension of a basic screening design by changing all signs; this assures that main effects will no longer be confounded with two-factor interactions in a combined analysis. Also called a *foldover*.

Repeat A repetition of an experimental run without reapplying the setup for that run. It is not an independent repetition of the run.

Replication A repetition of a run that is a complete and independent reapplication of the setup.

Response The output from the experiment. The dependent variable(s).

Robustness A condition in which changes in the setting of a factor do not have impact on the response variable; i.e., the response is not sensitive to noise in the factor settings.

Run A defined combination of the levels for all factors. Also called a *treatment*.

Screening designs Designs that provide information primarily on main effects, incurring the risks associated with confounding, in order to determine which of a large number of factors merit further study.

Sparsity The characteristic that main effects are generally larger than two-factor interactions while three-factor interactions can usually be ignored.

Treatment A defined combination of the levels for all factors. Also called a *run*.

t-test The statistical test used to assess the significance of the difference in averages.

Variance The square of the standard deviation.

Index

A
Aliases, 38–40, 109
Alpha risks, 13, 109
Analysis of effects, 7–16
    calculating effects, 7, 8–10
    decision limits, 7, 13–14
    graphing significant effects, 7, 14
    modeling significant effects, 7, 14–15
    Pareto charts, 7, 11, 15, 49, 59, 62, 110
    standard deviation of the effects, 7, 13
    standard deviation of the experiment, 7, 11–12
    t-statistic, 7, 13, 24, 66, 110
Analysis matrix, 109
Analysis of results, 3, 5
Analysis table for 1/2 fractional factorial with four factors, 93
Analysis table for 1/2 fractional with five factors, 97
Analysis table for 1/4 fractional with five factors, 95
Analysis table for 8-run Plackett-Burman design, 99
Analysis table for 8-run Plackett-Burman design with reflection, 101
Analysis table for 12-run Plackett-Burman design with reflection, 104
Analysis table for 16-run Plackett-Burman design, 105
Analysis table for 16-run Plackett-Burman design with reflection, 106
Analysis table for reflection of 8-run Plackett-Burman design, 100
Analysis tables for 12-run Plackett-Burman design, 102, 103
Analysis tables for two by three factorial, 90–91
Analysis with unreplicated experiments, 34–36
Appendix, 65–107
Averages, differences in, 8

B
Base designs, 46
Bias, 6, 16
Blocking, 7, 60, 61
Bond Strength (Example 1), 5–15
Brainstorming, 16, 61

C
Calculating effects, 7, 8–10
Calculators, 1, 12
Candidate process variables, 3, 5, 16
Cause-and-effect diagrams, 16
Cause-and-effect relationships, 1, 2
Center points, 24, 109
Chemical-Processing Yield (Example 3), 27–33
Coal-Processing Yield (Exercise 1), 23, 68–71
Common-cause variations, 2
Computer analysis, 60
    See also Software applications
Conflicts, 33
Confounding, 38–40, 50, 109
Confounding pattern for 8-run Plackett-Burman design, 98
Confounding pattern for 16-run Plackett-Burman design, 98
Control charting, 2
Control systems, 2
Corrosion Study (Exercise 2), 25, 72–75
Cost constraints, 7, 57, 60, 61, 62
Curvature, 7, 57

D
Decision limits (DL), 7, 13–14
Defined relationships, 39
Degrees of freedom (d.f.), 13, 36, 60–61, 109
    table for, 66
Dependent variables, 110
    See also Responses
Design of experiments (DOE), 1, 109
    basics of, 61–62
    definition, 2, 6
    fundamental concept of, 8
    logic steps in planning and implementing, 5, 16
    objectives, 3, 31
    reasons not used, 62
DesignExpert (TM), 63
Design matrix for 1/2 fractional factorial with four factors, 92
Design matrix for 1/4 fractional factorial with five factors, 94
Design matrix for 20-run Plackett-Burman, 107
Design matrix, 109

Design table for 1/2 fractional with five factors, 96
d.f. *See* Degrees of freedom
Diamond factors, 33
Differences in averages, 8
Distortions, 6, 16, 60
DL. *See* Decision limits
Documenting the experiment, 5, 62
DOE. *See* Design of experiments
Dummy factors, 60

E
Effect heredity. *See* Heredity
Effects
    definition of, 8
    importance of, 8
Effect sparsity. *See* Sparsity
Eight-Run Plackett-Burman Design with Seven Factors (Example 4), 40-49
Eight-step analytical procedure. *See* Analysis of effects
Eighty/twenty rule, 15
Evolutionary operation (EVOP), 57, 63
EVOP. *See* Evolutionary operation
Example 1: Bond Strength, 5-15
Example 2: Water Absorption in Paper Stock, 17–23
Example 3: Chemical-Processing Yield, 27–33
Example 4: Eight-Run Plackett-Burman Design with Seven Factors, 40–49
Example 5: Moldability Analysis, 51–55
Exercise 1: Coal-Processing Yield, 23, 68–71
Exercise 2: Corrosion Study, 25, 72–75
Exercise 3: Ink Transfer, 33–34, 76–80
Exercise 4: Nail Pull Study, 50–51, 81–84
Exercise 5: The Lawn Fanatics, 55–56, 85–89
Experimental design. *See* Design of experiments
Experimental errors, 11, 16, 24, 109
Experiments with three factors. *See* Three factor experiments
Experiments with two factors. *See* Two factor experiments
Extraneous variables, 5, 16
Eyeball assessment, 53

F
Factorial designs
    analysis tables for two by three design, 90–91
    comparison of number of runs in factorial and screening designs, 37
    fractional, 39, 57, 65, 109
    full, 6, 57, 109
Factors, 1, 6, 109
    high or plus sign, 6, 7, 22
    low or minus sign, 6, 7, 22
    number of, 21–22, 60, 61
    See also Appendix; Variables

False effects, 16
Flowcharts, 16
Fold-over designs, 46–49, 109, 110
Formal tests of significant effects, 59
Fractional factorial designs, 39, 57, 65, 109
F-statistic, 36, 109
    F-table for, 36, 67
Full-factorial designs, 6, 57, 109

G
Geometric designs, 39–40, 60, 109
Golden factors, 33
Grand average of the experiment, 59
Graphing significant effects, 7, 14
Graphs, 62

H
Heredity, 40, 45, 54, 109
Hierarchy rule, 14, 109

I
Identical interactions, 38
Independent variables, 6
Inertia, 62
Inherent variation, 2, 6, 7, 11
Ink Transfer (Exercise 3), 33–34, 76–80
Input variables, 1
Interactions, 2, 9, 14, 21, 60, 109
Interpolation procedures, 14–15
Interpretation of results, 3

L
L-18 designs, 57
Large designs, 57
The Lawn Fanatics (Exercise 5), 55–56, 85–89
Levels, 5, 6, 7, 61, 109
Linear relationships, 7, 23
Logic steps, 5, 16
Logic tools, 1
Loss of information, 59–60
Lower specification limits, 2

M
Main effects, 8, 21, 46, 50, 60, 110
Management support, 62
Manpower, 62
Manufacturing, 62
Market share, 62
Missing data, 59–60
Mixture designs, 57, 63
Modeling significant effects, 7, 14–15
Moldability Analysis (Example 5), 51–55
Myths, 62

N

Nail Pull Study (Exercise 4), 50–51, 81–84
NCSS, 63
Nongeometric designs, 40, 60, 110
Nonlinear relationships, 7, 57
    guidelines for, 24–25
    procedures for, 23–24
Nuisance-factor batches, 7
Number Cruncher (TM), 63

O

OFAT. *See* One-factor-at-a-time
One-factor-at-a-time (OFAT), 2
One-tailed tests, 32
Optimization designs, 50, 63
Outliers, 33, 49, 59, 110
Output measures, 1, 3, 5
    See also Response variables

P

Pareto charts, 7, 11, 15, 49, 59, 62, 110
Performance of the experiment, 3, 5, 61–62
Physical constraints, 7, 61
Plackett-Burman designs, 39–40, 51–55, 57, 60, 110
    See also Appendix
Predicted (pred) responses, 35
Prediction equation, 14
Procedures for experimentation, 3
Process capability, 2
Process noise, 2
Process variables, 1, 16
Process variation, 2
Proposing next study, 5, 16

Q

Quadratic relationships. *See* Nonlinear relationships
Qualitative factors, 6
Quantitative factors, 6, 16, 24

R

Randomization, 6, 16, 59, 60, 61
Random variation, 13
Raw material, 60
Refining designs, 110
Reflection designs, 46–49, 59, 60, 61, 110
Repeats, 6, 110
Replications, 6, 16, 49, 59, 60–61, 110
Reporting conclusions, 62
Residual (res) responses, 36
Responses, 1, 7, 110
Response surface designs, 57
Response variables, 5
    See also Output measures
Results, 7

Robustness, 2, 110
Runs, 6, 16, 60, 110

S

Sales, 62
Screening designs, 16, 37–57, 60, 61, 110
    aliases, 38–40, 109
    comparison of number of runs in factorial and screening designs, 37
    confounding, 38–40, 50, 109
    definition, 37
    Example 4: Eight-Run Plackett-Burman Design with Seven Factors, 40–49
    Example 5: Moldability Analysis, 51–55
    Exercise 4: Nail Pull Study, 50–51, 81–84
    Exercise 5: The Lawn Fanatics, 55–56, 85–89
    heredity and sparsity, 40, 45, 54, 109, 110
    intent of, 39
Sigma, 11
Significant effects, 2
    formal tests of, 59
    graphing, 7, 14
    modeling, 7, 14–15
    number of, 59
Software applications, 1, 36, 62–63
    for fractional factorial designs, 40
    for large designs, 57
    and missing data, 60
    See also Computer analysis
Sparsity, 40, 45, 54, 110
SPC. *See* Statistical process control
Special-cause variation, 2, 16, 33, 49, 59
Spreadsheets, 21–23, 28, 36, 63
Standard deviation of the effects, 7, 13
Standard deviation of the experiment, 7, 11–12
STAT-EASE, Inc., 63
Statistical control, 16
Statistical process control (SPC), 1, 2
    sigma, 11
Statistics, 2
Systems approach, 2

T

Table for F-values, 67
Table of t-values, 66
Taguchi designs, 57, 63
Technical experts, 62
Tests of significance, 49
Test statistics, 2
Three factor experiments, 27–36, 57
    Example 3: Chemical-Processing Yield, 27–33
    Exercise 3: Ink Transfer, 33–34, 76–80
Training, 62
Treatments, 6, 16, 60, 110
t-statistic, 7, 13, 24, 110
    table for, 66

Twelve-run Plackett-Burman design, 51–55
Two factor experiments, 5–25
    Example 1: Bond Strength, 5–15
    Example 2: Water Absorption in Paper Stock, 17–23
    Exercise 1: Coal-Processing Yield, 23, 68–71
    Exercise 2: Corrosion Study, 25, 72–75
Two-factor interactions, 46
Two-tailed test, 32

U

Unknown or uncontrolled factors, 6
Unreplicated experiment analysis, 34–36
Upper specification limits, 2

V

Variables, 1
    importance of control, 3
    See also Factors
Variance formula, 11–12, 110
Variation, 2, 16
Variation analysis, 49
    guidelines, 33
    procedure, 31–33

W

Water Absorption in Paper Stock (Example 2), 17–23